A VISTAS OF SCIENCE BOOK

Experimentation and Measurement

By W. J. YOUDEN
Statistical Consultant
Applied Mathematics Division
National Bureau of Standards
Washington, D.C.

ADVISORY GROUP

Science Educator: J. STANLEY MARSHALL
Professor of Science Education
The Florida State University
Tallahassee, Florida

Research Scientist: T. A. WILLKE
Mathematical Statistician
National Bureau of Standards
Washington, D.C.

Science Teacher: DALE E. GERSTER
Bladensburg High School
Bladensburg, Maryland

Produced by the NATIONAL SCIENCE TEACHERS AS-
SOCIATION in cooperation with the NATIONAL BUREAU
OF STANDARDS. Published by SCHOLASTIC BOOK SERV-
ICES, A Division of Scholastic Magazines, Inc., New York,
New York.

Copyright © 1962 by the National Science Teachers Association, Inc.
All rights reserved. Published by Scholastic Book Services.

2nd printing .... May 1962

Manufactured in the U.S.A.

Library of Congress Catalog Card Number: 62-13537

NATIONAL SCIENCE TEACHERS ASSOCIATION

1201 Sixteenth Street, N.W., Washington 6, D.C.

A Department of the National Education Association
An Affiliate of the American Association for the Advancement of Science

VISTAS OF SCIENCE BOOKS

Exciting discoveries, major breakthroughs, and important new applications continue to widen our vistas of science, engineering, and technology, producing an impact on society which touches every individual. These expanding frontiers inspire large numbers of our youth to consider careers in science, and thoughtful persons of all ages seek a better understanding of the scientific-technological society in which we live. As a result, there is an insatiable demand for current, accurate scientific information. To fill this pressing need, the National Science Teachers Association has conceived and developed the VISTAS OF SCIENCE series. In so doing, NSTA advances its own central purpose: improvement of the teaching of science. In addition, VISTAS books provide scientific background for those who would be well-informed, responsible citizens.

Designed at the request of students and teachers, the series is produced under the guidance of an experienced Advisory Board (see facing page). Each book is concerned with a specific science area, such as spacecraft, oceanography, astronomy, the cell, genetics, measurement.

Three types of information characterize VISTAS OF SCIENCE books: presentation of subject matter, research frontiers and methods, and student activities. VISTAS are science resource and enrichment literature that is sound and challenging. Written for junior and senior high school students, the VISTAS are of interest and value to teachers and other adults as well.

Sponsors of VISTAS books include governmental agencies, professional societies, and industrial organizations. These groups help to support the program financially and cooperate to insure that the manuscripts are developed as objective, authentic explorations of particular science areas.

The authors are scientists, science writers, and classroom teachers of outstanding experience, background, and communicative skills. Often a combination of authors develops the manuscript — a scientist or science writer prepares the main text, and a classroom teacher prepares the student activity section.

An advisory group, composed of a research scientist, a science educator, and a science teacher, assists in the planning and reviewing of each manuscript. These advisers represent three groups vitally concerned with providing sound education in science.

The VISTAS OF SCIENCE books are made available by the publisher through its school book-club programs and by individual purchase. One copy of each VISTA is provided free to all school science clubs enrolled in NSTA's Future Scientists of America organization.

The VISTAS program, through the cooperative efforts of authors, advisory groups, sponsors, NSTA, and the publisher, makes possible low-cost, versatile science libraries in the science classrooms of the nation's schools and encourages individual students to plan home science libraries as well.

The suggestions and comments of students, teachers, and other readers of VISTAS OF SCIENCE books will be welcome.

Robert H. Carleton
Executive Secretary, NSTA

Marjorie Gardner
Director, VISTAS OF SCIENCE

CONTENTS

One approach to the topic "measurement" would be the historical and factual. The curious early units of measurement and their use would make an interesting beginning. The development of modern systems of measurement and some of the spectacular examples of very precise measurement would also make a good story.

I have chosen an entirely different approach. Most of those who make measurements are almost completely absorbed in the answer they are trying to get. Often this answer is needed to throw light on some tentative hypothesis or idea. Thus the interest of the experimenter is concentrated on his subject, which may be some special field of chemistry or physics or other science.

Correspondingly, most research workers have little interest in measurements except as they serve the purpose of supplying needed information. The work of making measurements is all too often a tiresome and exacting task that stands between the research worker and the verification or disproving of his thinking on some special problem. It would seem ideal to many research workers if they had only to push a button to get the needed data.

The experimenter soon learns, however, that measurements are subject to errors. Errors in measurement tend to obscure the truth or to mislead the experimenter. Accordingly, the experimenter seeks methods to make the errors in his measurements so small that they will not lead him to incorrect answers to scientific questions.

In the era of great battleships there used to be a continuous struggle between the makers of armor plate and the gunmakers who sought to construct guns that would send projectiles through the latest effort of the armor plate manufacturers. There is a somewhat similar contest in science. The instrument makers continually devise improved instruments and the scientists continually undertake problems that require more and more accurate measurements. Today, the requirements for accuracy in measurements often exceed our ability to meet them. One consequence of this obstacle to scientific research has been a growing interest in measurement as a special field of research in itself. Perhaps we are not getting all we can out of our measurements. Indeed, there may be ways to use presently available instruments to make the improved measurements that might be expected from better, but still unavailable, instruments.

We know now that there are "laws of measurement" just as fascinating as the laws of science. We are beginning to put these laws to work for us. These laws help us understand the errors in measurements, and they help us detect and remove sources of error. They provide us with the means for drawing objective, unbiased conclusions from data. They tell us how much data will probably be needed. Today, many great research establishments have on their staffs several specialists in the theory of measurements. There are not nearly enough of these specialists to meet the demand for them.

Thus I have thought it more useful to make this book an elementary introduction to the laws of measurements. But the approach is not an abstract discussion of measurements, instead it depends upon getting you to make measurements and, by observing collections of measurements, to discover for yourself some of the properties of measurements. The idea is to learn something about measurement that will be useful — no matter what is being measured. Some hint is given of the devices that scientists and measurements specialists use to get more out of the available equipment. If you understand something about the laws of measurements, you may be able to get the answers to your own research problems with half the usual amount of work. No young scientist can afford to pass up a topic that may double his scientific achievements.

—W. J. YOUDEN

1. INTRODUCTION

The plan

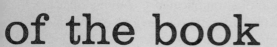

of the book

Measurements are made to answer questions such as: How long is this object? How heavy is it? How much chlorine is there in this water?

In order to make measurements we need suitable units of measurement. When we ask the length of an object, we expect an answer that will tell us how many inches, or how many millimeters it is from one end of the object to the other end.

We need some way to find out how many times a unit quantity, such as a millimeter, is contained in the length of a particular object. Rarely will a unit of length go into the length of the object a whole number of times. Almost always our answer will be, "so many units plus some fractional part of a unit." If a quantitative measurement of length is to be trusted, we must take great care that the unit we choose is invariable and suitable for the task. We must also have devised a method of measuring the object with this unit.

Some measurements require only a simple procedure and little equipment. The apparatus may be no more than a scale marked off in the desired units. It is easy to measure the width of a table by using a meter stick marked off in centimeters and millimeters. The air temperature of a room is found by looking at a thermometer and reading the position of the mercury on the scale. The pressure in an automobile tire is found by applying a tire gauge to the valve and looking at the scale to read the pounds of air pressure per square inch of surface in the tire.

When the proper instrument is available and used carefully, many measurements require no more than a careful reading of a scale. On the other hand, most scientific measurements involve elaborate equipment and a complicated technique of using it.

If a chemist wants to determine the amount of chlorine in a material, he may perform a fairly lengthy sequence of operations. He must first weigh out a sample of the material and record the weight. The sample must be treated with an acid that will dissolve out all of the chlorine. Any insoluble residue must be filtered off to obtain a clear solution, and the filter paper must be washed carefully with excess acid to make sure that none of the chlorine is left behind.

It then may be necessary to adjust either the acid concentration or the volume of the solution — or both — before adding a second reagent to precipitate the chlorine. The usual reagent is silver nitrate. Enough must be added to precipitate all the chlorine as insoluble silver chloride. This precipitate of silver chloride is separated from the acid by filtering the suspension through a crucible with a porous bottom.

Before doing this, however, it will be necessary to weigh the crucible, making sure that it is dry. The precipitate collected in the crucible should be then washed with distilled water to remove all traces of reagent and dried. The weight of the empty crucible subtracted from the weight of the crucible and the precipitate gives the weight of the silver chloride.

By using the atomic weights of silver and chlorine, the proportion of chlorine in the silver chloride molecule can be determined. The weight of silver chloride precipitate multiplied by this proportion gives the weight of chlorine in the precipitate. This, of course, is also the weight of the chlorine in the original sample. The weight of chlorine divided by the weight of the sample and multiplied by 100 gives the per cent of chlorine in the sample, thus completing the determination of chlorine.

The Errors in Measurements

If we consider that each weighing (sample, empty crucible, and crucible plus precipitate) is a measurement, we see that three measurements are necessary to measure the amount of chlorine in the material. This sketch of the analytical procedure reveals that there are several steps, all of which must be taken with great care. If the silver chloride precipitate is not carefully washed, the silver chloride may be contaminated and appear too heavy. If the precipitate is not transferred completely to the crucible, some may be lost. None of these steps can be carried out so that they are absolutely free of error. For example, since the silver chloride is very slightly soluble, some of the chloride will not be precipitated. This results in error.

Evidently a measurement is subject to many sources of error, some of which may make the measurement too large, while others may tend to make the measurement too small. It is the aim of the experimenter to keep these sources of error as small as possible. They cannot be reduced to zero. Thus in this or any measurement procedure, the task remains to try to find out how large an error there may be. For this reason, information about the sources of errors in measurements is indispensable.

In order to decide which one of two materials contains the larger amount of chlorine, we need accurate measurements. If the difference in chlorine content between the materials is small and the measurement is subject to large error, the wrong material may be selected as the one having the larger amount of chlorine. There also may be an alternative procedure for determining chlorine content. How can we know which procedure is the more accurate unless the errors in the measurements have been carefully studied?

Making Measurements

The best way to find out about some of the difficulties in making measurements is to make measurements. Much of this book will be devoted to making measurements — to trying to find out something about the sources of errors in measurements and how they may be detected and minimized.

The second chapter is an easy one. It goes a little more into detail about the importance of making good measurements and tells us something about the role of measurements in our everyday life and in business and commerce. In the third chapter we undertake a measurement that involves no more equipment than a book and a millimeter scale. Everyone who reads this book should try making several repetitions of the measurement described there. We will examine 96 similar measurements made by a class of girls. Such a large collection of measurements poses the problem of finding some convenient and precise method of describing the collection. Perhaps we can find some number to represent the whole collection and some other number that will represent the errors in the measurements.

When you have made the measurements described in Chapter 3, you will have completed your first exploration of the world of measurement. It will be natural for you to wonder if some of the things that you have learned on the first exploration apply to other parts of the measurement world.

Scientific measurements are often time consuming and require special skills. In Chapter 4, we will examine the reports of other explorers in the world of measurement. You may compare their records with the results you found in order to see if there is anything in common. I hope you will be delighted to find that the things you have observed about your own measurements also apply to scientific and engineering data.

Mapping the Land of Measurement

One of the primary tasks of all explorers — and scientists are explorers — is to prepare a map of an unknown region. Such a map will serve as a valuable guide to all subsequent travelers. The measurements made by countless researchers have been studied by mathematicians and much of the world of measurements has been mapped out. Not all of it, by any means, but certainly enough so that young scientists will be greatly helped by the existing maps. So Chapter 5 may be likened to a simplified map of regions already explored. Even this simplified map may be something of a puzzle to you at first.

Remember, Chapter 5, like a map, is something to be consulted and to serve as a guide. Yet people get lost even when they have maps. Don't be surprised if you get lost. By the time you have made some more measurements, which amounts to exploration, you will begin to understand the map better and will be able to use it more intelligently.

The rest of the book concerns some other journeys in the land of measurement. Now that you have a map, you are a little better equipped. The next set of measurements that you can undertake yourselves requires the construction of a small instrument. Most measurements do involve instruments. It is a good idea to construct the instrument yourself.

Next we undertake an exploration that requires a team of four working together. These easy measurements will reveal how much can be learned from a very few measurements.

Then comes a chapter which discusses in a brief manner another important problem confronting most investigators. We cannot measure everything. We cannot put a rain gauge in every square mile of the country. The location of the rain gauges in use constitutes a *sample* of all possible sites. Similarly, we cannot test all the steel bars produced by a steel mill. If the test involved loading each bar with weights until it broke, we would have none left to use in construction. So a sample of bars must be tested to supply the information about the strength of all the bars in that particular batch. There is an example in this chapter that shows something about the sampling problem.

The final chapter describes a more complicated measurement and the construction and testing of a piece of equipment. All research involves some kind of new problem and the possibility of requiring new apparatus. Once you have constructed a piece of equipment and made some measurements with it, your respect for the achievements of the research worker will increase. Making measurements that will be useful to scientists is an exacting task. Many measurements are difficult to make. For this reason we must make the very best interpretation of the measurements that we do get. It is one of the primary purposes of this book to increase your skill in interpretation of experiments.

2. Why we need measurements

A measurement is always expressed as a multiple of some unit quantity. Most of us take for granted the existence of the units we use; their names form an indispensable part of our vocabulary. Recall how often you hear or use the following words: degree Fahrenheit; inch, foot, and mile; ounce, pound, and ton; pint, quart, and gallon; volt, ampere, and kilowatt hours; second, minute, and day. Manufacturing and many other commercial activities are immensely helped by the general acceptance of standard units of measurement. In 1875, there was a conference in Paris at which the United States and eighteen other countries signed a treaty and established an International Bureau of Weights and Measures. Figure 1 shows a picture of the International Bureau in France.

Numbers and Units

The system of units set up by the International Bureau is based on the meter and kilogram instead of the yard and pound. The metric system is used in almost all scientific work. Without a system of standard units, scientists from different countries would be greatly handicapped in the exchange of scientific information. The task of defining units still goes on. The problem is not as easy as it might seem. Certain units may be chosen arbitrarily; for example, length and mass. After four or five units are established in this way, it turns out that scientific laws set up certain mathematical relations so that other units — density, for example — are derived from the initial set of units.

Obvious also is the need of a number system. Very likely the evolution of number systems and the making of measurements were closely related. Long ago even very primitive men must have made observations that were counts of a number of objects, such as the day's catch of fish, or the numerical strength of an army. Because they involve whole numbers, counts are unique. With care, they can be made without error; whereas measurements cannot be made exactly.

Air temperature or room temperature, although reported as 52°F., does not change by steps of one degree. Since a report of 51° or 53° probably would not alter our choice of clothing, a report to the nearest whole degree is satisfactory for weather reports. However, a moment's thought reveals that the temperature scale is continuous; any decimal fraction of a degree is possible. When two thermometers are placed side by side, careful inspection nearly always shows a difference between the readings. This opens up the whole problem of making accurate measurements.

We need measurements to buy clothes, yard goods, and carpets. The heights of people, tides, floods, airplanes, mountains, and satellites are important, but involve quite different procedures of measurements and the choice of appropriate units. For one or another reason we are interested in the weights of babies (and adults), drugs, jewels, automobiles, rockets, ships, coins, astronomical bodies, and atoms, to mention only a few. Here, too, quite different methods of measurements — and units — are needed, depending on the magnitude of the weight and on the accessibility of the object.

Significance of Small Differences

The measurement of the age of objects taken from excavations of bygone civilizations requires painstaking measurements of the relative abundance of certain stable and radioactive isotopes of carbon; C-12 and C-14 are most commonly used. Estimates of age obtained by carbon dating have a known probable error of several decades.

Another method of measuring the age of burial mounds makes use of pieces of obsidian tools or ornaments found in them. Over the centuries a very thin skin of material — thinner than most paper — forms on the surface. The thickness of this skin, which accumulates at a known rate, increases with age and provides an entirely independent measure of the age to compare with the carbon-14 estimate. Here time is estimated by measuring a very small distance.

Suppose we wish to arrange some ancient objects in a series of ever-increasing age. Our success in getting the objects in the correct order depends on two things: the difference in ages between objects and the uncertainty in the estimate of the ages. Both are involved in attaining the correct order of age.

If the least difference in age between two objects is a century and the estimate of the age of any object is not in error by more than forty years, we will get the objects in the correct order. Inevitably the study of ancient civilizations leads to an effort to get the correct order of age even when the differences in age are quite small. The uncertainty in the measurement of the age places a definite limitation on the dating of archeological materials.

The detection of small differences in respect to some property is a major problem in science and industry. Two or more companies may submit samples of a material to a prospective purchaser. Naturally the purchaser will want first of all to make sure that the quality of the material he buys meets his requirements. Secondly, he will want to select the best mate-

Figure 1.   Measurement standards for the world are maintained

rial, other things, such as cost, being equal.

When we buy a gold object that is stated to be 14 carats *fine*, this means that the gold should constitute 14/24 of the weight. We accept this claim because we know that various official agencies occasionally take specimens for chemical analysis to verify the gold content.

An inaccurate method of analysis may lead to an erroneous conclusion. Assuming that the error is in technique and not some constant error in the scales or chemicals used, the chemical analysis is equally likely to be too high as it is to be too low. If all the items were exactly 14 carats, then chemical analysis would show half of them to be below the specified gold content. Thus an article that is actually 14 carats fine might be unjustly rejected, or an article below the required

in Paris at the International Bureau of Weights and Measurements.

content may be mistakenly accepted. A little thought will show that if the error in the analysis is large, the manufacturer of the article must make the gold content considerably more than 14/24 if he wishes to insure acceptance of nearly all the items tested.

There are two ways around this dilemma. The manufacturer may purposely increase the gold content above the specified level. This is an expensive solution and the manufacturer must pass on this increased cost. Alternatively, the parties concerned may agree upon a certain permissible *tolerance* or departure from the specified gold content. Inasmuch as the gold content cannot be determined without some uncertainty, it appears reasonable to make allowance for this uncertainty. How large a tolerance should be set? This will depend primarily on the accuracy of the chemical analysis. The point is that, besides the problem of devising a method for the analysis of gold articles, there is the equally important problem of determining the sources of error and size of error of the method of analysis. This is a recurrent problem of measurement, regardless of the material or phenomenon being measured.

There may be some who feel that small differences are unimportant because, for example, the gold article will give acceptable service even if it is slightly below 14 carats. But small differences may be important for a number of reasons. If one variety of wheat yields just one per cent more grain than another variety, the difference may be unimportant to a small farmer. But added up for the whole of the United States this small difference would mean at least ten million more bushels of wheat to feed a hungry world.

Sometimes a small difference has tremendous scientific consequences. Our atmosphere is about 80 per cent nitrogen. Chemists can remove the oxygen, carbon dioxide, and moisture. At one time the residual gas was believed to consist solely of nitrogen. There is an interesting chemical, ammonium nitrite, NH_4NO_2. This chemical can be prepared in a very pure form.

When heated, ammonium nitrite decomposes to give nitrogen (N_2) and water (H_2O). Now pure nitrogen, whether obtained from air or by the decomposition of NH_4NO_2, should have identical chemical and physical properties. In 1890, a British scientist, Lord Rayleigh, undertook a study in which he compared nitrogen obtained from the air with nitrogen released by heating ammonium nitrite. He wanted to compare the densities of the two gases; that is, their weights per unit of volume. He did this by filling a bulb of carefully determined volume with each gas in turn under standard conditions: sea level pressure at $0°$ centigrade. The weight of the bulb when full minus its weight when the nitrogen was exhausted gave the weight of the nitrogen. One measurement of the weight of atmospheric nitrogen gave 2.31001 grams. Another measurement on nitrogen from ammonium nitrite gave 2.29849 grams. The difference, 0.01152, is small. Lord Rayleigh was faced with a problem: was the difference a measurement error or was there a real difference in the densities? On the basis of existing chemical knowledge there should have been no difference in densities. Several additional measurements were made with each gas, and Lord Rayleigh concluded that his data were convincing evidence that the observed small difference in densities was in excess of the experimental errors of measurement and therefore actually existed.

There now arose the intriguing scientific problem of finding a reason for the observed difference in density. Further study finally led Lord Rayleigh to believe that the nitrogen from the air contained some hitherto unknown gas or gases that were heavier than nitrogen, and which had not been removed by the means to remove the other known gases. Proceeding on this assumption, he soon isolated the gaseous element argon. Then followed the discovery of the whole family of the rare gases, the existence of which had not even been suspected. The small difference in densities, carefully evaluated as not accidental, led to a scientific discovery of major importance.

Tremendous efforts are made to improve our techniques of

making measurements, for who knows what other exciting discoveries still lie hidden behind small differences. Only when we know what are the sources of error in our measurements can we set proper tolerances, evaluate small differences, and estimate the accuracy of our measurements of physical constants. The study of measurements has shown that there are certain properties common to all measurements; thus certain mathematical laws apply to all measurements regardless of what it is that is measured. In the following chapters we will find out some of these properties and how to use them in the interpretation of experimental data. First, we must make some measurements so we can experience first hand what a measurement is.

3. Measurements in experimentation

T HE object of every scientific experiment is to answer some question of interest to a scientist. Usually the answer comes out in units of a system of measurement. When a measurement has been made the scientist trusts the numerical result and uses it in his work, if the measurement apparatus and technique are adequate. An important question occurs to us right away. How do we know that the measurement apparatus and technique are adequate? We need rules of some kind that will help us to pass judgment on our measurements. Later on we will become acquainted with some of these checks on measurements.

Our immediate task is to make some measurements. The common measurements made every day, such as reading a thermometer, differ in a very important respect from scientific measurements. Generally we read the thermometer to the nearest whole degree, and that is quite good enough for our purposes. If the marks are one degree apart, a glance is enough to choose the nearest graduation. If the interval between adjacent marks is two degrees, we are likely to be satisfied with a reading to the nearest even number. If the end of the mercury is approximately midway between two marks, we may report to the nearest degree, and that will be an odd-numbered degree.

The Knack of Estimating

Fever thermometers cover only a small range of temperature. Each whole degree is divided into fifths by four smaller marks between the whole degree graduation marks. The fever thermometer is generally read to the nearest mark. We get readings like 98.6°, 99.8°, or 100.2° F. As the fever rises, readings are taken more carefully and the readings may be estimated between marks, so that you may record 102.3°. Notice that body temperature can easily be read to an extra decimal place over the readings made for room temperatures. This is possible because the scale has been expanded.

Examine a room thermometer. The graduation marks are approximately one sixteenth of an inch apart. The mercury may end at a mark or anywhere in between two adjacent marks. It is easy to select a position midway between two marks. Positions one quarter and three quarters of the way from one mark to the next mark are also fairly easy to locate. Usually we do not make use of such fine subdivisions because our daily needs do not require them. In making scientific measurements, it is standard practice to estimate positions in steps of one tenth of the interval. Suppose the end of the mercury is between the 70 and 71 degree mark. You may feel a little uncertain whether the mercury ends at 70.7° or at 70.8°. Never mind, put down one or the other. Practice will give you confidence. Experts may estimate the position on a scale to one twentieth of a scale interval. Very often a small magnifying glass is used as an aid in making these readings.

Here is an example of a scientific problem that requires precise temperature readings. Suppose that you collect some rain water and determine that its freezing point is 32°F. Now measure out a quart of the rain water sample and add one ounce of table sugar. Place a portion of this solution in a freezing-brine mixture of ice and table salt, stirring it all the while. Ice will not begin to appear until the temperature has dropped to a little more than 0.29°F., below the temperature at which your original sample begins to turn to ice.

From this simple experiment you can see that freezing points can be used to determine whether or not a solvent is pure. The depressions of the freezing point produced by dissolving substances in solvents have long been a subject of study. In these studies temperatures are usually read to at least one thousandth of a degree, by means of special thermometers. The position of the mercury is estimated to a tenth of the interval between marks which are 0.01 of a degree apart. Very exact temperature measurements taken just as the liquid begins to freeze can be used to detect the presence of minute amounts of impurity.

In scientific work the knack of estimating tenths of a division on scales and instrument dials becomes almost automatic. The way to acquire this ability is to get some practice. We will now undertake an experiment that will quickly reveal your skill in reading subdivisions of a scale interval. The inquiry that we are to undertake is to measure the thickness of the paper used in one of your textbooks. Although a single sheet of paper is much thinner than the smallest scale divisions on a ruler, a simple procedure will make it possible for you to determine the thickness of the paper quite accurately. The procedure consists of reading the thickness of a stack of sheets of the paper and dividing the reading by the number of sheets in the stack. Simple as this procedure appears, we will find that it reveals quite a lot about measurements.

Prepare a form for recording the measurements. Strangely enough, careful measurements are sometimes so poorly recorded even by professional scientists that not even the experimenter can interpret them sometime later. A form suitable for this experiment with spaces to enter four separate observations is shown in Table 1. Note that the form identifies the observer, the source of the data, and the date of the experiment or observation. These are characteristics of useful forms.

Choose a book without inserts of special paper. First open the book near the beginning and also near the end. Hold the stack of pages together. This is the stack whose thickness we will measure with a ruler marked off in centimeters and millimeters. We will estimate tenths of a millimeter.

There are a number of details to observe even in this simple

Table 1. Example of a form for keeping a record of measurements. Experiment to measure thickness of sheets of paper

Observer: B. G.   Date: March 16, 1961

Book title and author: MODERN CHEMISTRY, Dull, Metcalfe and Williams

obser-vation number	page number at front	at back	pages in stack	sheets in stack	stack thickness mm.	thickness per sheet mm.
1	67	387	320	160	12.4	0.0775
2	41	459	418	209	16.4	0.0785
3	23	549	526	263	20.0	0.0760
4	35	521	486	243	18.5	0.0761
					Total	0.3081
					Average	0.0770

[Data taken by a student at Immaculata High School, Washington, D.C.]

experiment. Read the numbers on the first page of the stack and that on the page facing the last page of the stack. Both numbers should be odd. The difference between these two numbers is equal to the number of pages in the stack. This difference must always be an even number. Each separate sheet accounts for two pages, so divide the number of pages by two to get the number of sheets. Enter these data on the record form.

Pinch the stack firmly between thumb and fingers and lay the scale across the edge of the stack. Measure the thickness of the stack and record the reading. The stack will usually be between one and two centimeters thick; i.e., between 10 and 20 mm. (millimeters). Try to estimate tenths of a millimeter. If this seems too hard at first, at least estimate to the nearest one fourth (0.25) of a millimeter. Record these readings as decimals. For example, record 14 and an estimated one fourth of a division as 14.25.

Measurements Do Not Always Agree

After you have made the first measurement, close the book. Then reopen it at a new place and record the new data. Make at least four measurements. Now divide the reading of the thickness by the number of sheets in the stack. The quotient gives the thickness of one sheet of paper as a decimal part of a millimeter. When this division is made for each measurement, you will certainly find small differences among the quotients. You have discovered for yourself that measurements made on the *same* thing do not agree perfectly. To be sure, the number of sheets was changed from one measurement to the next. But that does not explain the disagreement in the answers. Certainly a stack of 200 sheets should be just twice as thick as a stack of 100 sheets. When the stack thickness is divided by the number of sheets we should always get the thickness of a single sheet.

There are two major reasons for the disagreement among the answers. First, you may pinch some of the stacks more tightly than others. You could arrange to place the stack on a table top and place a flatiron on the stack. This would provide a uniform pressure for every measurement. The second major reason is your inexpertness in reading the scale and estimating between the scale marks. Undoubtedly this is the more important of the two reasons for getting different answers. The whole reason for insisting on closing the book was to make sure the number of sheets was changed each time. You knew the thickness would change and expected to get a change in your scale reading. Unless you are very good in mental arithmetic you could not predict your second scale reading.

Suppose, however, that you were able to do the necessary proportion in your head. If you knew in advance what the second scale reading should be to make it check with your first result, this would inevitably influence your second reading. Such influence would rob the second reading of the indispensable independence that would make it worthy of being called a measurement.

It may be argued that all four answers listed in Table 1 agree in the first decimal place. Clearly, all the answers are a little less than 0.08 mm. Thus any one of the results rounded off would give us this answer. Why take four readings?

Just to take a practical everyday reason, consider the paper business. Although paper in bulk is sold by weight, most users are also concerned with the thickness of paper and size of the sheet. Thick paper will mean fewer sheets of a given size per unit weight paid for. A variation of as little as 0.01 mm. — the difference between 0.08 mm. and 0.09 mm. — would reduce the number of sheets by more than ten per cent. We need to know the answer to one or two more decimal places. In a situation like this, it is usual to obtain the average of several readings. You should note, however, that although repetition alone doesn't insure accuracy, it does help us locate errors.

Many people seem to feel that there is some magic in the repetition of measurements and that if a measurement is repeated frequently enough the final result will approach a "true" value. This is what scientists mean by accuracy.

Suppose that you were in science class and that the next two people to come into your classroom were a girl five feet ten inches tall and a boy five feet nine inches tall. Let each of the 30 students already in the class measure the two new arrivals to the nearest foot. The answer for both is six feet. Has repeated measurement improved the accuracy?

Suppose that the hypothesis to be tested was that girls are taller than boys. This time the boy and the girl were each measured 30 times with a ruler that read to 1/100 of an inch. Could we conclude that the repeated measurements really supported the hypothesis? The point is that repeated measurements alone do not insure accuracy. However, if a set of measurements on the same thing vary widely among themselves we begin to suspect our instruments and procedures. This is of value if we are ever to achieve a reasonable accuracy.

Paper thickness is so important in commerce that the American Society for Testing Materials has a recommended procedure for measuring the thickness of paper. A standard pressure is put on the stack and a highly precise instrument called a micrometer is used to measure the thickness. Even then the results show a scatter, but farther out in the decimal places. Improved instruments do not remove the disagreement among answers. In fact the more sensitive the apparatus, the more surely is the variation among repeat measurements revealed. Only if a very coarse unit of measurement is used does the disagreement disappear. For example, if you report height only to the nearest whole meter practically all adult men will be two meters tall.

It used to be a common practice among experimenters to pick out the largest and smallest among the measurements and report these along with the average. More often today the difference between the largest and smallest measurement is reported together with the average from all the measurements. This difference between the maximum and minimum values is called the *range* of the group. The range gives some notion of the variation among the measurements. A small range, or a range which is a small percentage of the average, gives us more confidence in the average. Although the range does reveal the skill of the operator, it has the disadvantage of depending on the number of measurements in the group. The same operator usually will find his average range for groups of ten measurements to be about 1.5 times as large as the range he gets for groups of four measurements. So the number of measurements in the group must always be kept in mind.

Averages, Ranges, and Scatter

The data for operator B. G. have been given in complete detail in Table 1. This operator was one of a class of 24 girls, all of whom made four measurements on the same book. This larger collection of data will reveal still more about measurements. The measurements made by these girls are tabulated in Table 2, which shows the computed thickness for each trial. The details of pages and millimeters have been omitted. Most of the girls did not estimate tenths of a millimeter but did read to the nearest quarter millimeter. Two or three had readings only to the nearest whole millimeter. A gross misreading of the scale was evidently made by girl U on her last trial. This value has been excluded from the average and no range entered for this student.

The remaining 23 ranges vary widely. This does not necessarily mean that some girls were better than others in reading the scale. Even if all girls had the same skill, the range may vary severalfold when it is based on just four measurements. Of course, if this class of girls repeated the measurements and the very small ranges and very large ranges were produced by the same girls as before, this would indicate that some girls can repeat their measurements better than other girls. One way to summarize the results is to give the average thickness, 0.07709, and the average of the 23 ranges, 0.00483. The last two places of decimals are quite uncertain for both these averages. Indeed, a mathematician would say we cannot be sure that the second seven in 0.07709 is correct. Additional decimal places are always carried when making computations and these should be entered in the data record.

Table 2. Tabulation of 96 measurements of paper thickness made by 24 girls.

girl		thickness in mm.			average	range	+	−
A	.0757	.0762	.0769	.0746	.0758	.0023	0	4
B	.0808	.0793	.0781	.0821	.0801	.0040	4	0
C	.0811	.0772	.0770	.0756	.0777	.0055	2	2
D	.0655	.0683	.0714	.0746	.0700	.0091	0	4
E	.0741	.0710	.0748	.0711	.0728	.0038	0	4
F	.0756	.0772	.0776	.0759	.0766	.0020	2	2
G*	.0775	.0785	.0760	.0761	.0770	.0025	2	2
H	.0747	.0765	.0735	.0776	.0756	.0041	1	3
I	.0719	.0762	.0802	.0713	.0749	.0089	1	3
J	.0734	.0833	.0833	.0783	.0796	.0099	3	1
K	.0755	.0740	.0714	.0743	.0738	.0041	0	4
L	.0788	.0817	.0794	.0766	.0791	.0051	3	1
M	.0731	.0716	.0726	.0714	.0722	.0017	0	4
N	.0833	.0794	.0783	.0788	.0800	.0050	4	0
O	.0767	.0775	.0765	.0793	.0775	.0028	2	2
P	.0787	.0798	.0864	.0817	.0816	.0077	4	0
Q	.0784	.0799	.0789	.0802	.0794	.0018	4	0
R	.0784	.0820	.0796	.0818	.0804	.0036	4	0
S	.0830	.0796	.0778	.0767	.0793	.0063	3	1
T	.0741	.0680	.0733	.0723	.0719	.0061	0	4
U	.0759	.0766	.0772	.0466**	.0766	—	1	3
V	.0810	.0812	.0789	.0776	.0797	.0036	4	0
W	.0777	.0759	.0795	.0790	.0780	.0036	3	1
X	.0784	.0786	.0797	.0859	.0806	.0075	4	0

Total for 95 measurements   = 7.3239

Average for 95 measurements = 0.07709

Average for 23 ranges   = 0.00483

*G is the same student, B.G., reported in Table 1.

**The last measurement made by student U is .0466. This appears to be a mistake as it is little more than half as large as the other measurements. This measurement is omitted from the collection and the total and average computed from the remaining 95 measurements.

Notice that the number of sheets in the stack involves three significant figures. The quotient or thickness per sheet is therefore carried out to three significant figures. To use only two figures would be equivalent to rounding off the number of sheets before dividing. Failure to carry enough figures tends to conceal the variation in the data.

Condensing Our Data

Tabulated results, as shown in Table 2, look like a sea of numbers. There is a way to bring out the essential characteristics of such collections of many measurements on the same thing. Paradoxically we may condense the data into more compact form and at the same time get a better picture of the collection. The smallest of the 95 results is 0.0655 mm. and the largest is 0.0864 mm. The range for these 95 results is therefore 0.0209. Suppose we form a series of intervals to cover this range. We may start out at 0.0650 mm. and make each interval equal to 0.0020 mm. The size of the interval should be small enough so that at least six intervals will be needed. If there are many measurements there should be more intervals than with few measurements. Table 3 shows eleven intervals that completely cover the whole range of values.

The intervals are written down in a column. Then each of the values in Table 2 (except the apparent mistake) is placed in its proper interval by making a pen stroke opposite the interval class. The actual values are, in effect, replaced by the mid-values of the interval class to which they have been assigned. The slight change made by using the mid-values of the intervals is of no consequence. Indeed, some values are slightly increased and others decreased. Much of the effect therefore cancels out.

Now we are beginning to get some order in our sea of numbers. The mass of individual items in the data now have been replaced by the eleven different mid-values along with the number of measurements assigned to each mid-value. The last column shows the product of each mid-value by the number of measurements that go with it. The total for this column, 7.32925, is close to the total, 7.3239 (Table 2), of the actual measurements. The averages obtained by dividing each total by 95 are 0.07715 and 0.07709. The difference is quite unimportant.

Table 3. Retabulation of data in Table 2

measurement interval	number of measurements in this interval		mid-value of interval	no. times mid-value
.0650 - .0669	/	1	.06595	.06595
.0670 - .0689	//	2	.06795	.13590
.0690 - .0709		0	.06995	–
.0710 - .0729	/// /// /// /	10	.07195	.71950
.0730 - .0749	/// /// /// ///	12	.07395	.88740
.0750 - .0769	/// /// /// /// /// ///	18	.07595	1.36710
.0770 - .0789	/// /// /// /// /// /// /// ///	24	.07795	1.87080
.0790 - .0809	/// /// /// /// //	14	.07995	1.11930
.0810 - .0829	/// /// //	8	.08195	.65560
.0830 - .0849	/// /	4	.08395	.33580
.0850 - .0869	//	2	.08595	.17190
	Total	95		7.32925
			Average	0.07715

Grouping measurements into intervals is standard practice. The presentation of the data is more concise. Furthermore a glance at the check marks made opposite the intervals in Table 3 tells us something about the data. We see that the interval containing the average contains more measurements than any other — it is called the *modal interval.* Intervals on either side of the modal interval have fewer measurements in them. The number in each interval falls off sharply near the end intervals. Apparently measurements that differ considerably from the average are relatively scarce. This is an encouraging thought for experimenters. Obviously, however, there is a chance of getting one of these scarce measurements. Experimenters are naturally much interested in knowing what the risk is of getting a measurement quite distant from the average.

Often the counts of the measurements in the intervals are shown graphically. One way to do this is by means of a *histogram* as shown in Figure 2. To make this histogram, the 11 intervals were marked off as equal segments on a horizontal line. A suitable scale is laid off on a vertical line to designate the number of measurements in each interval. Horizontal bars are drawn at the proper heights and connected as shown. The general form of this histogram, the intervals, and the number in each interval, tell the expert just about everything that the actual measurements would.

We have seen one histogram and obtained some idea of the way this collection of measurements is distributed around an

Figure 2. Histogram for 95 measurements of paper thickness

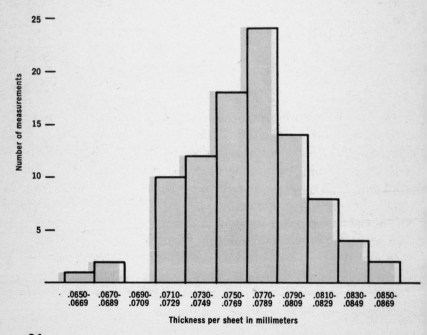

average. In Chapter 4 several different collections of measurements are represented by histograms. You will then be able to observe that in many collections of measurements there are similarities in the distributions regardless of the objects being measured. This fact has been of crucial importance in the development of the laws of measurement.

Let's return to our measurements of paper thicknesses and investigate some of the properties of this collection. The measurements in the collection should meet certain requirements. One of these requirements is that each of the four measurements made by a student should be a really independent measurement. By that we mean that no measurement is influenced by any following measurement. Another requirement is that all participants should be equally skillful. If some measurements were made by a skilled person and some by a novice, we should hesitate to combine both collections. Rather we should make a separate histogram for each individual. We would expect the measurements made by the skillful one to stay closer to the average. His histogram might be narrow and tall when compared with the histogram for the novice. The readings made by the novice might be expected to show a greater scatter. Histograms can provide a quick appraisal of the data and the technique of the measurer.

Four measurements are too few to rate any individual. Nevertheless, the availability of 24 individuals makes it possible to explore still another property of these data. If we think about the measurement procedure, we see that it is reasonable to assume that any given measurement had an equal chance of being either larger or smaller than the average. In any particular measurement the pressure on the stack could equally well have been either more or less than the average pressure. The scale reading may have erred on the generous side or on the skimpy side. If these considerations apply, we would expect a symmetrical histogram. Our histogram does show a fair degree of symmetry.

Before we conclude that the requirements for putting all the measurements into one collection have been fully satisfied, we must carefully examine the data. The reason we changed the number of pages for each measurement was to avoid influencing later readings by preceding readings. If we happened to get too large a reading on the first measurement, this should not have had the effect of making subsequent readings too large. We are assuming, of course, that the pressure applied to the stack varied with each measurement, and that the reading of the scale was sometimes too large and sometimes too small. It also seems reasonable to assume that there is a 50-50 chance of any one measurement being above or below the average value. Is this true of the measurements made by the girls in this science class?

It is conceivable, of course, that a particular individual always squeezes the paper very tightly and in consequence always gets lower readings than the average for the class. Another person might always tend to read the scale in a way to get high readings. If this state of affairs exists, then we might expect that all readings made by a particular individual would tend to be either higher or lower than the average, rather than splitting 50-50.

Let us think about a set of four measurements in which each measurement is independent and has the same chance to be more than the average as it has to be less than the average. What kind of results could be expected by anyone making the four measurements? One of five things must happen: All four will be above the average, three above and one below, two above and two below, one above and three below, all four below.

Our first impulse is to regard a result in which all four measurements are above (or below) the average as an unlikely event. The chance that a single measurement will be either high or low is 50-50, just as it is to get heads or tails with a single coin toss. As an illustration, suppose a cent, a nickel, a dime, and a quarter are tossed together. The probabilities of four heads, three heads, two heads, one head, or no heads are easily obtained. The pos-

Table 4. Possible ways four coins may fall when they are tossed together.

	cent	nickel	dime	quarter
One way to get no heads	T	T	T	T
Four ways to get only one head	H	T	T	T
	T	H	T	T
	T	T	H	T
	T	T	T	H
Six ways to get only two heads	H	H	T	T
	H	T	H	T
	H	T	T	H
	T	H	H	T
	T	H	T	H
	T	T	H	H
Four ways to get only three heads	H	H	H	T
	H	H	T	H
	H	T	H	H
	T	H	H	H
One way to get four heads	H	H	H	H

sible ways the four coins might fall are enumerated in Table 4.

There are just sixteen different ways in which the coins may fall. We may easily calculate our chances of getting no heads, one head, two, three, or four heads. For example, we find there is only one way to get four heads — the chance is 1 in 16. Remember that this calculation assumes that a tossed coin is equally likely to fall heads as it is tails. Incidentally, the chances are not altered if four cents are used, as you can determine for yourselves by trying it out. The mathematical experts among the readers will know that $(H + T)^4 = H^4 + 4H^3T + 6H^2T^2 + 4HT^3 + T^4$. Observe that the coefficients 1, 4, 6, 4, 1 correspond to the counts shown in Table 4. Some of you may be inclined to find out whether or not this relationship holds if three, five, or n coins are tossed instead of four coins.

Let's now see how the results from tossing four coins can serve as a useful model in examining the collection of measurements made on the thickness of paper. If — as in the case of heads or tails, when coins are tossed — high or low readings are equally likely, we conclude that there is 1 chance in 16 of getting four high readings and 1 chance in 16 of getting four low readings. There are 4 chances in 16 of getting just one high reading and an equal chance of getting just three high readings. Finally there are 6 chances in 16 of getting two high readings and two low readings.

Now to apply this model to the entire collection of 24 sets of four measurements each, we can multiply each of the coefficients on the preceding page by 1.5 ($24/16 = 1.5$). This will give us the expected frequencies of highs and lows for 24 sets of four measurements as shown in the third line of Table 5.

We must not expect that these theoretical frequencies are going to turn up exactly every time. You can try tossing four coins 24 times and recording what you get. There will be small departures from theory, but you may confidently expect that in most of the 24 trials you will get a mixture of heads and tails showing on the four coins.

The last two columns in Table 2 are headed by a plus and by a minus sign. In those columns the individual readings are compared with the average of all the readings, 0.07709, to determine whether they are above (plus) or below (minus) the average. Note that girl A had four readings all below the average, so four is entered in the minus column and zero in the plus column. Girl B's readings are just the reverse, all four are above the average. Girl C had two above and two below. We next count up the frequencies for the various combinations, and find them to be 6, 3, 4, 4, and 7 respectively. These numbers are entered in the fourth line of Table 5.

When we examine these frequencies a surprising thing confronts us. We find far too many girls with measurements either all above or all below the average. In fact there are 13 of these against an expected three. This disparity is far too great to be accidental. Evidently our assumed model does not fit the facts.

Table 5. Five ways to place 24 sets of four measurements with reference to the average.

High readings (Above the average)	4	3	2	1	0	
Low readings (Below the average)	0	1	2	3	4	TOTAL
Expected frequency for 24 measurements	1.5	6	9	6	1.5	24
Observed frequency (from our data)	7	4	4	3	6	24

The hope of complete independence for the readings has not been realized. It seems that if the first reading was high, subsequent readings also tended to be high. The same holds true if the first reading happened to be low. Evidently many of these girls had a particular way of measuring that persisted throughout all four measurements. We see that for many of these girls agreement of the four measurements with each other does not tell the whole story. All four measurements may be quite high or quite low. We sometimes say that such individuals are subject to biases.

Bias—a Major Consideration

Once a scientist or measurement specialist detects or even suspects that his readings are subject to a bias, he tries to take steps to locate the bias and to correct his measurement procedure. The goal is to reduce bias as far as possible. Experience shows that only rarely can biases be completely eliminated. We can be quite sure in this case that some of the girls have rather marked biases and this complicates the interpretation of the data. Nevertheless, since there are nearly as many girls with plus biases as those with negative biases, the histogram is still reasonably symmetrical.

One way to think about these measurements is to regard the

set of four measurements made by any one girl as having a certain scatter about her own average. Her average may be higher or lower than the class average; so we may think of the individual averages for all the girls as having a certain scatter about the class average. Even this simple measurement of the paper thickness reveals the complexity and problems of making useful measurements. A measurement that started out to be quite simple has, all of a sudden, become quite a complicated matter, indeed.

One more property of these data should be noted. Table 2 lists the average of the four measurements made by each girl. There are 23 of these averages (one girl's measurements were excluded). The largest average is 0.0816 and the smallest is 0.0700. The largest of the measurements, however, was 0.0864 and the smallest was 0.0655. Observe that the averages are not scattered over as wide a range as the individual measurements. This is a very important property for averages.

In this chapter we have used data collected in only a few minutes by a class of girls. Just by looking at the tabulation of 96 values in Table 2 we found that the measurements differed among themselves. A careful study of the measurements told us quite a lot more.

We have learned a concise and convenient way to present the data, and that a histogram based on the measurements gives a good picture of some of their properties. We also observed that averages show less scatter than individual measurements. And most interesting of all, perhaps, we were able to extract from these data evidence that many of the students had highly personal ways of making the measurement. This is important, for when we have located shortcoming in our ways of making measurement we are more likely to be successful in our attempts to improve our measurement techniques.

4. Typical

collections of

measurements

In the preceding chapter a careful study was made of 96 measurements of the thickness of paper used in a textbook. We learned how to condense the large number of measurements into a few classes with given mid-values. The mid-values together with the number in each class provided a concise summary of the measurements. This information was used to construct a histogram, which is a graphical picture of how the measurements are distributed around the average value of the measurements. In this chapter a number of collections of data will be given together with their histograms. We are going to look for some common pattern in the way measurements are distributed about the average.

The first example concerns 100 measurements to determine the amount of magnesium in different parts of a long rod of magnesium alloy. Chemists find it convenient to have on hand specimens of an alloy of known composition. Such specimens make it easy for the chemist to calibrate the equipment used in the analysis of metals and make sure that it is working properly.

In this example, an ingot of magnesium alloy was drawn into a rod about 100 meters long and with a square cross section about 4.5 centimeters on a side. The long rod was cut into 100 bars, each a meter long. Five of the bars were selected at random and a flat test piece about 1.2 centimeters thick was cut from each. These served as test specimens.

Table 6. Duplicate determinations of magnesium at 50 test points

position	bar 1		bar 5		bar 20		bar 50		bar 85	
1	0.076	0.067	0.069	0.066	0.073	0.070	0.073	0.063	0.070	0.069
2	0.071	0.066	0.071	0.062	0.069	0.068	0.075	0.069	0.066	0.064
3	0.070	0.065	0.068	0.066	0.068	0.066	0.069	0.067	0.068	0.063
4	0.067	0.066	0.071	0.066	0.069	0.067	0.072	0.068	0.068	0.064
5	0.071	0.065	0.066	0.065	0.070	0.068	0.069	0.066	0.064	0.087
6	0.065	0.067	0.068	0.066	0.070	0.065	0.069	0.065	0.070	0.065
7	0.067	0.067	0.071	0.067	0.065	0.068	0.072	0.082	0.069	0.064
8	0.071	0.067	0.069	0.067	0.067	0.069	0.063	0.063	0.067	0.064
9	0.066	0.063	0.070	0.065	0.067	0.073	0.069	0.066	0.069	0.068
10	0.068	0.068	0.068	0.068	0.064	0.068	0.069	0.066	0.067	0.069

On each of the five specimens ten test points were located in the pattern shown in Figure 3. This gave 50 spots in all. Two determinations of the magnesium content were made at each spot.

The total collection of 100 determinations is shown in Table 6. The determinations range from 0.062 to 0.082 per cent of magnesium. One unit in the last place was used as the interval for drawing the histogram in Figure 4. To avoid crowding the scale the intervals are labeled 62, 63, . . . instead of 0.062, 0.063. . . . On the histogram the lone high reading of 0.082 shows up like a sort thumb. How can we account for it?

Perhaps the analyst misread his instrument. That seems more likely than to

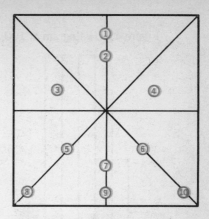

Figure 3. Pattern for locating test points on cross section of alloy bar.

assume the existence of a single isolated spot much higher in magnesium than the 49 other spots. The safest guide to choosing between these alternatives would be to repeat that analysis. In fact a duplicate analysis of that spot was made and gave the value 0.072. The duplicates differ by 0.010.

We may get some more help in this situation by examining the other 49 differences between the duplicates. The analyst ran all 50 spots once and then made a set of repeat determinations. When the results of the second set are subtracted from the first results as shown in Table 7, an interesting state of affairs is revealed. Plus differences predominate. There are 40 plus differences and only ten negative differences. As a rule, the entire second set seems to be lower than the first set. One might assume that under normal conditions there would be no reason to expect the second measurement on a spot to be smaller than the first one. It would be more reasonable to expect that it would be a toss up as to which would be the larger, the first result or the second. Again it is like tossing a coin 50 times and observing the number of heads obtained. Theory predicts that we should expect to get close to 25 heads.

A surplus or a deficit of seven heads would be rather rare.

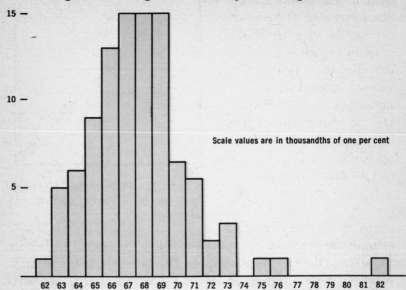

Figure 4. Histogram of 100 analyses for magnesium

Scale values are in thousandths of one per cent

A surplus or deficit of ten heads would be most unusual. A surplus of 15 heads would make most of us very suspicious about the fairness of the coin or the improper skill of the one tossing it. Thus we cannot ascribe this large preponderance of plus differences to mere chance. But how can we account for it? First of all, we note that when the second measurement is subtracted from the first measurement the average difference, taking account of the sign, is $+0.0022$. This together with the fact that the first entry for each pair was run as a group supplies the clue to the mystery.

Evidently something happened to the apparatus between doing the first group of 50 and the second 50 determinations. Apparently between the first and second runs there was some small shift in the apparatus or environment and this introduced a bias. As a result the apparatus began to give lower results.

The surprising fact is that the suspect high value of 0.082 was made in the second run and is larger, not smaller, than its com-

Table 7. Difference between duplicate determinations of magnesium. First duplicate minus the second duplicate. Entries show the number of plus and minus differences. The four zero differences have been divided between plus and minus.

difference	plus	minus
0.000	2	2
0.001	5	—
0.002	8	3
0.003	7	2
0.004	5	1
0.005	8	—
0.006	2	1
0.007	—	—
0.008	—	—
0.009	2	—
0.010	1	1*
Total	40	10

*The asterisk identifies the difference associated with the suspect measurement 0.082.

panion first determination of 0.072. This large difference is ten units off in the wrong direction from the +2.2 average difference. (We multiplied the average difference by 1000 to drop the zeros.) The 50 differences listed in Table 7 are exhibited as the histogram at the top of Figure 5. The large difference of ten units of the wrong sign is crosshatched. It stands apart and furthest removed from the average difference of +2.2. It should be evident by this time that not only is the 0.082 an out-of-line high value but also that it is responsible for the largest difference of any of the 50 pairs — and of the wrong sign. Surely these facts justify the deletion of this result. The single first determination is left to represent the seventh test point on bar No. 50.

Before you conclude that we are being fussy about a small error, remember we are showing that mistakes do occur. Sometimes mistakes tip the scales of judgment one way or the other. We are all prone to lapses in technique. Therefore we need to be prepared to search for questionable values.

Figure 5.   Upper histogram shows distribution of the 50 differences between the duplicates shown in Table 7.

Lower histogram shows distribution of the 50 differences found between randomly selected the pairs drawn from Table 6. See also Table 8.

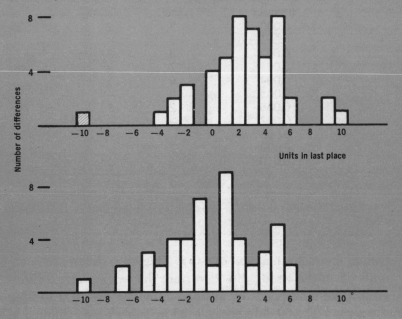

Units in last place

Histograms and the Error of Measurement

The analysts doing this work very wisely planned to do the determinations in two complete groups. Suppose that the first group of 50 determinations had consisted of duplicate analyses of the first five test points on each bar. The consequence would have been to make test points six through ten appear to be lower in magnesium than test points one through five. Thus there would have been real doubt about the homogeneity of the alloy sample. However, the plan used by the analysts kept the comparison between test points fair. Each of the 50 test spots is represented in the first 50 determinations and again in the second 50.

We still have not answered the question in the mind of the experimenters who made these determinations. We want to know if the stock of material, as represented by these 50 spots actually tested, is acceptably uniform to serve as a standard for testing other samples of magnesium alloy. To answer this question we must know whether the differences between spots can be ascribed to the measurement error or whether they represent real differences in magnesium content. If the values of differences found between determinations on different spots are similar to the values of the differences found between duplicates on the same spot, we would be satisfied with the uniformity. There may be minute differences in concentration of magnesium at the various test spots. Clearly they are not important if the variation in concentration is considerably less than the error of measurement, i.e., the error inherent in the technique.

The question is, how do we determine which of these alternatives is the case? A direct test may be made in the following way. Write the numbers 1 to 50 on 50 cards and shuffle them well. Cut the deck of cards in two places and read the two numbers. These correspond with a pair of test spots. Copy one determination for each number, but make sure that both are either first or second determinations. Can you explain why? Repeat this process 50 times. How many possible pairs are there to choose from?

Suppose you cut the cards and turn up numbers 19 and 33. Look back to Table 6 and read the values opposite these numbers, *making sure you take readings from the same set of 50.* If you select the first run of 50 (the five left-hand columns), the values would be 0.070 (for number 19) and 0.069 (for number 33). Subtract the second value from the first to obtain a difference of +0.001. Replace the cards, shuffle, and cut again. This time you cut numbers 30 and 46. Still using the first set of 50 determinations, you find the values 0.064 and 0.070. This time the difference is —0.006. Continue until you have 50 differences in any pattern. If duplicates turn up, cut again.

Table 8. Differences between fifty pairs of determinations picked at random from fifty spots subject to the condition that both members of the pair are from the same group of fifty. Difference is equal to the first member of the pair minus the second member of the pair.

difference	plus	minus
0.000	1	1
0.001	9	7
0.002	4	4
0.003	2	4
0.004	3	2
0.005	5	3
0.006	2	—
0.007	—	2
0.008	—	—
0.009	—	—
0.010	—	1
Total	26	24

When this game was actually tried, it produced the results shown in Table 8. These results are represented in the histogram at the bottom of Figure 5. If we compare the two histograms, we see that they are spread over nearly the same width and are remarkably similar in form. The histogram for the duplicates is displaced to the right 2.2 units as a consequence of the shift in readings between the first and second groups of measurements. This shift does not change the form or width of the diagram.

The displacement in the top histogram exists because the average for the first set of 50 results is 2.2 units higher than the average for the second set of 50 results. The duplicate spots selected for the bottom histogram were always chosen from the same 50 results (either the first 50 or the second 50) so if we take the sign into account, the average difference should be zero.

You may verify this statement by subtracting 2.2 units from each of the 50 differences listed in Table 7 and making a third histogram which shows the differences corrected for the shift. So, regardless of the shift, the width of the top histogram truly represents the error of the measurement. Now, the plan of work has eliminated the shift error from the comparison of test points with the result that the shift error can also be excluded.

Duplicate determinations were made on the same spot so there could have been no actual change in magnesium concentration, only changes as a result of measurement error. The differences arising from comparing different parts of the bars are shown in the lower histogram of Figure 5. This histogram is about the same width as the top histogram for duplicates. If, on the other hand, there had been a marked variation in magnesium concentration in the rod, the differences between determinations from different locations would have exceeded the duplicate differences. Hence we see that since the determinations made on different parts of the bar agree about as well as repeat determinations made on one spot, we conclude, therefore, that the bar is homogeneous enough for our purpose.

The experiment just discussed is comparatively simple as experiments go. Nevertheless the interpretation of the data required a good deal of care. What have we learned that will guide us in future measurements? First, we see that no matter what we measure or how we measure it, we will always require a knowledge of the measurement error itself. We have learned that shifts in the apparatus may occur and that we can protect the experiment from such shifts by a proper plan of work. Visible, too, is the general similarity in shape between the histogram for the magnesium analyses and for the measurements on paper thickness. We also devised quite a useful method of testing whether the concentration of magnesium varied more from test point to test point than could be accounted for by the error in the measurement itself. This procedure can be adapted to a large variety of measurements.

The similarity noticed in the general shape of the histograms for paper measurements and for spectrographic chemical analyses was not accidental. Whether the measurements are approximate or very precise the shape persists. Nor does it matter

49

whether the measurements are in the fields of physics, chemistry, biology, or engineering.

The histogram in Figure 6 was made from the results obtained

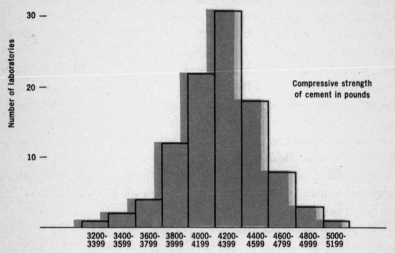

Figure 6. Histogram for cement tests reported by 102 laboratories.

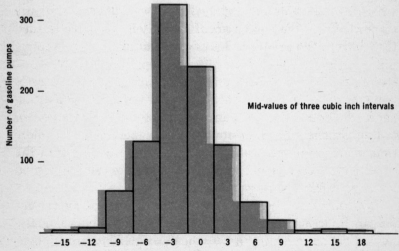

Figure 7. Histogram showing cubic inches of error in 923 gasoline pumps tested by inspectors.

when 102 laboratories tested samples of the same batch of cement. This was done to track down the source of disagreement between tests made by different laboratories. From the histogram made from the data it is clear that a few laboratories were chiefly responsible for the extremely high and low results.

All states and many cities provide for the regular inspection of gasoline pumps to ensure that the amount of gasoline delivered stays within the legal tolerance for five gallons. From these tests a large amount of data becomes available. Remember that the manufacturer first adjusts the pump so that it will pass inspection. Naturally the station owner does not want to deliver more gasoline than he is paid for. A small loss on each transaction over a year could be disastrous. However, the pump itself cannot be set without error; nor can the inspector who checks the pump make measurements without error.

The scatter of the results exhibited by the histogram in Figure 7 reflects the combined uncertainty in setting and checking the pump. In this group of 923 pumps only 40 had an error greater than one per cent of the number of cubic inches (1155) in five gallons. This was the amount pumped out for these tests.

Some extremely fine measurements are displayed as histograms in Figure 8. These have to do with a redetermination of the gravitational constant g (the acceleration due to gravity) in Canada. The procedure involves timing the fall of a steel bar and measuring the speed with which it falls. To time the bar's fall, a light beam and photoelectric cells are used. As the bar drops, the light beam is reflected from precisely placed mirrors on the bar and signals the photoelectric cells. These precisely timed signals measure with great accuracy how fast the steel bar is falling. Thus the acceleration due to gravity is calculated.

The scientists, in the course of exploring possible sources of error in their work, made 32 measurements with each of two different bars. Inspection of the histograms shows a good deal of overlap of the results with the two bars. The average for bar number 1 is 980.6139 cm./sec.^{2}, and for bar number 2 is 980.6124

Figure 8. Two groups of 32 measurements of the acceleration due to gravity at Ottawa, Canada.

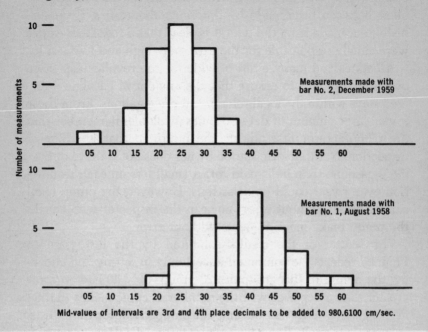

Mid-values of intervals are 3rd and 4th place decimals to be added to 980.6100 cm/sec.

cm./sec.² The difference between these averages is about 15 parts in almost ten million. The bars agree to almost one part in a million. Small though this difference is, the evidence leads to the conclusion that there is a real difference between the results with the two bars. The typical shape of the histogram appears again in these superb measurements.

Your author, in his younger days, was a research chemist in a biological research institute. The experiments carried out there often required substantial collections of plants. But even when plants are started at the same time, in the same soil, and grown on the same bench in the same greenhouse, they vary a great deal in their growth. The biologists always asked the greenhouse staff to start more plants than would be needed. When an experiment was started, the biologist would pick out a supply of plants with very closely the same growth. The extremely small and large plants would be discarded. Usually some plants were used for controls and others given various experimental treatments. It was an advantage to begin each experiment with a stock of fairly uniform plants, so that control and treated plants would start off even.

The Normal Law of Error

In experiments of this type great care must be taken to avoid personal bias in selecting experimental and control plants. If extra-good plants are unconsciously assigned to a particular treatment the experiment certainly will be biased.

Suppose there are 50 plants to be divided into five groups of ten plants. One group will be used as controls; the other four groups are to be assigned to treatments 1, 2, 3, and 4. How shall the groups be formed? Prepare a deck of 50 cards. Mark C on ten of the cards, mark 1 on ten more, and so on. Shuffle the deck thoroughly. Arrange the 50 plants in a row and deal out

Figure 9. Draw a line over the plants in photo below, and you have a curve of the normal law of error.

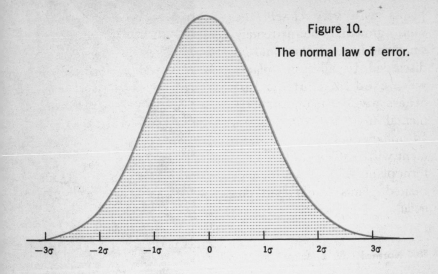

Figure 10.

The normal law of error.

-3σ -2σ -1σ 0 1σ 2σ 3σ

the deck of cards as you move down the row. Chance, not unconscious personal bias, determines the outcome. There are many variations of this technique of assigning the plants to the control and treatment groups. Unless precautions are taken at this stage, the most searching study of the data may still lead to incorrect conclusions. We will never know for sure whether the performance of the best group is really the result of the experimental treatment, or because it was favored in the assignment of the plants.

One day, just before the stock of plants was about to be culled of the very small and very large ones, a simple experiment was tried. The smallest plant in the lot was singled out and placed at the left end of the bench. Then the largest plant was placed at the opposite right end. There was room for 15 plants between. A selection of 15 plants was made that provided a regular increase in size from the smallest to the largest plant. This took a little while, but finally a satisfactory "scale" was established along the edge of the bench. A summer high school assistant was assigned the task of "matching up" the remaining plants with the 17 plants used to establish a scale of size. The pots were placed so that lines of plants of the same size extended back toward the center of the bench. When the sorting was over, what do you think the arrangement of the plants looked like? A picture of the arrangement is shown as Figure 9. The assistant, who had never heard of histograms, had arranged the plants in the familiar shape we have been finding for our histograms of paper measurements, magnesium determinations, and measurements of g.

If we drew a curved line over the histograms that we have examined so far, we would find that the curves would be symmetrical about their mid-points and have long tails on either side. A curve of this type is shown in Figure 10. The equation which determines the height of this curve and its extent on either side of the mid-point is known as the *normal law of error*.

Only two constants are needed to determine the height and width of this curve. These two quantities, which make it possible to construct a curve representing a histogram, can be calculated from any collection of data. We seldom go to the trouble of actually constructing the curve. Generally, we make direct use of the two constants which, together with statistical tables, are sufficient to interpret most collections of data. The equation for the constants is one of the great discoveries of science. Its importance in bringing meaning to collections of observations can hardly be overestimated.

Your author has a small printing press for a hobby. He set in type his opinion of the importance of the normal law of error.

THE
NORMAL
LAW OF ERROR
STANDS OUT IN THE
EXPERIENCE OF MANKIND
AS ONE OF THE BROADEST
GENERALIZATIONS OF NATURAL
PHILOSOPHY ◆ IT SERVES AS THE
GUIDING INSTRUMENT IN RESEARCHES
IN THE PHYSICAL AND SOCIAL SCIENCES AND
IN MEDICINE AGRICULTURE AND ENGINEERING ◆
IT IS AN INDISPENSABLE TOOL FOR THE ANALYSIS AND THE
INTERPRETATION OF THE BASIC DATA OBTAINED BY OBSERVATION AND EXPERIMENT

5. Mathematics of measurement

B Y this time we should be familiar with the construction and appearance of histograms. We now follow up the idea that a particular type of equation can be written for the curve that can be made to fit these histograms. This equation, called the normal law of error, is an exponential one of the form

$$y = \frac{1}{\sigma \sqrt{2\pi}} \, e^{-\frac{(x - \mu)^2}{2\sigma^2}}$$

Do not let the formidable appearance of this equation alarm you; y and x are simply the y axis and x axis coordinates. You are already familiar with the constant, π, the ratio of the circumference of a circle to its diameter. The constant e with a value of 2.7183 is the base of Napierian, or natural logarithms.

The other constants, μ (mu) and σ (sigma) depend on the experiment. For a given experiment μ and σ are fixed numbers, but their values are generally not known. In fact, the main purpose of an experiment is often to find out what these values really are. In that case the data from the experiment is used to provide estimates of their value.

Estimates of these two quantities are given the symbols m and s. Together with the mathematical tables for the values of x and y based on the equation above, these estimates are used universally both for routine measurements and the interpretation of research data.

One of the histograms used to represent the data obtained in the study of the acceleration due to gravity is shown in Figure 11. Superimposed on the histogram is the graph of the normal law of error constructed or "fitted" to the histogram by using the estimates of μ and σ calculated from the data. I hope you will agree that the curve is a good approximation to the outline of the histogram. The successful fitting of the normal error curve to the data using just two numbers justifies finding what these two numbers are and how to calculate them.

We know how to calculate one of them. This is the estimate, m, which is our old friend the arithmetic average of the data under a different name. This estimate is about as close as we can come to finding the true value of the constant μ in the question.

Recall the measurements you made on the thickness of paper. It seems reasonable, does it not, to assume that there is a real but unknown value for the thickness of the paper? The value of the constant is a quantity that is made up of the unknown true thickness of the paper plus any biases arising from using an imperfect scale in an individual manner. The distinction may

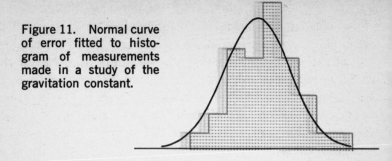

Figure 11. Normal curve of error fitted to histogram of measurements made in a study of the gravitation constant.

seem unnecessary but it is important to understand that μ is not the "true" thickness of the paper. Rather, μ is the value that the average, m, tends to approximate more and more closely as the number of measurements is increased. Of course, we hope that μ is very close to the true value. Nevertheless μ is utterly dependent on the error of the scale and on any bias in using the scale. We found that two individuals — each using his own scale — obtained averages that disagree substantially no matter how many measurements are made. Each individual apparently has his own μ. The true value for the thickness is indeed an elusive quantity.

The Estimate of Sigma

The second quantity calculated from the data is an estimate of σ, the *standard deviation*. This estimate of σ is given the symbol s; it determines the width of the normal error curve.

Turn back to Figure 10 and examine the curve sketched there. You will observe that if we drew a vertical line through the highest point on the curve, the curve would be symmetrical about it. The vertical line or the high point of the curve represents the value of the average for the data. It is customary to use this vertical line as a reference point for marking off multiples of the standard deviation to the right and to the left. You will also see that when we have proceeded as much as three standard deviations on either side of the center line, the curve has dropped to about one per cent of its height at the center. Table 9 gives the ordinates (y values) for the curve at k multiples of the standard deviation on either side of the mean. The table also gives the fraction of the area inclosed by any chosen pair of ordinates. Suppose that we take the location of the average to be the zero point, and we erect an ordinate at -1 standard deviation and another at $+1$ standard deviation. By referring to Table 9 we see that 68.27 per cent of the total area under the curve is included between these two ordinates. Ordinates erected at plus two standard deviations and minus two standard deviations include 95.45 per cent of the area. Similarly, ordinates at 2.57 standard deviations inclose 99 per cent of the area. We have observed how the curve approximates the outline of a histogram. Histograms are based on the number of measurements found in the chosen intervals. So we may use the above percentages (and others taken from Table 9) to obtain an indication of the expected per cent of the meas-

Table 9. Ordinates and areas for the normal curve of error		
x values given in multiples, k, of the standard deviation	y values given in multiples of 1/standard deviation	per cent of area included between ordinates at $-k\sigma$ and $+k\sigma$
± 0.00	0.3989	0.0000
0.25	0.3867	0.1974
0.50	0.3521	0.3829
0.75	0.3011	0.5467
1.00	0.2420	0.6827
1.25	0.1826	0.7887
1.50	0.1295	0.8664
1.75	0.0863	0.9199
2.00	0.0540	0.9545
2.25	0.0317	0.9756
2.50	0.0175	0.9876
2.75	0.0091	0.9940
3.00	0.0044	0.9973

urements that will fall within any given interval on the horizontal scale expressed in terms of the standard deviation.

Calculating s

In order to express an interval on the horizontal scale in terms of s, the estimate of σ, we must first calculate the value of s in terms of the units actually employed for the measurements. The formula for s is:

$$s = \sqrt{\frac{\Sigma d^2}{n - 1}}$$

Capital sigma, Σ, indicates a sum — in this case — of d^2, the squares of differences obtained by subtracting the average m for each measurement. Suppose we represent n measurements by the symbols $x_1, x_2, x_3, \ldots, x_n$. Then the average is found by adding the measurements and dividing the sum by n.

$$\text{Average} = m = \frac{x_1 + x_2 + x_3 + \ldots + x_n}{n} = \frac{\Sigma x}{n}$$

Finding s is simply a matter of substitution, as shown in Table 10.

But, in spite of the simple arithmetic, we are leading up to a truly remarkable generalization. We claim that in most sets of measurements, and without regard to what it is that is being measured, about two out of three measurements will differ from the average by less than one standard deviation. Similarly, about 19 measurements out of 20 will be within two standard deviations of the average. Only one measurement in one hundred can be expected to depart from the average by more than 2.57 standard deviations. Furthermore, these statements apply to the most varied sorts of measurements, whether they are precise or approximate. This property of the normal error curve is of great value for the interpretation of data.

Table 10. Calculating s, the estimate of the standard deviation

measurement	average	difference	square of difference
x_1	m	$x_1 - m = d_1$	d_1^2
x_2	m	$x_2 - m = d_2$	d_2^2
x_3	m	$x_3 - m = d_3$	d_3^2
•	•	•	•
x_n	m	$x_n - m = d_n$	d_n^2

Sum of squared differences $= \Sigma d^2$

$$s = \sqrt{\frac{\Sigma d^2}{n-1}}$$

We will illustrate the calculation of the standard deviation by using five made-up measurements.

measurement	measurement minus average	square of difference from average
27	5	25
26	4	16
23	1	1
19	−3	9
15	−7	49
Total 110	0	100
Average 22		

Now we can substitute the values in the equation and solve for s. The sum of the squares of the deviations from the average is 100. Divide this sum by one less than the number of measurements $(5 - 1 = 4)$ and obtain 25 as the quotient. The square root of 25 gives 5 as s, the estimate of the standard deviation.

The differences between each of the measurements and their average can now be expressed in multiples of s. All we do is divide by 5, i.e., by the estimate of the standard deviation. These differences become $1.0s$, $0.8s$, $0.2s$, $-0.6s$, and $-1.4s$.

Three of these five differences are less than one standard deviation and that is as good a check on the theoretical 68.27 per cent as can be obtained with so few measurements.

We cannot say with confidence that the actual value of σ is close to the estimate five when so few measurements have been used to calculate s. In order to get a really satisfactory estimate of s we like to have at least 30 measurements, but there are many experiments in which it is not possible to get this many. In dealing with problems when the measurements are few in number, however, a special allowance has to be made for the uncertainty of our estimate of s.

It is not actually necessary to fit the curve to the data. Recall the 95 measurements made on the thickness of paper used in a book. If m and s are determined for this collection, the expected number of measurements in each interval can be cal-

Figure 12. Solid line outlines actual histogram, dotted line outlines calculated histogram for measurements on the thickness of paper.

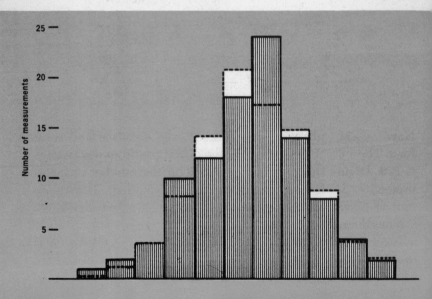

culated directly. These calculated values are shown in Figure 12 as a dotted-outline histogram superimposed over the actual histogram. The predicted values obtained by calculation conform fairly closely to the counts made on the measurements. Thus we see that either the curve or the histogram may be computed by using only two estimates calculated from the original data or from the data grouped into intervals. On the other hand, having s and m we can get a very good picture of the measurements without doing either.

Consider carefully what we have achieved in the way of condensing many measurements to a more useful form. First, the 95 measurements were sorted into eleven intervals. The mid-values of these intervals, together with the number of measurements in each interval, replaced the 95 measurements. This procedure in itself often makes for a very substantial reduction in the space required to report the data. The data may be reported either graphically as a histogram, or in brief tabular form of two columns; one column listing the mid-values of the intervals, the other showing the number of measurements for each mid-value. As a second step, the histogram or condensed table can be replaced by just two numbers, m and s, which permit us to reconstruct the original histogram, although the reconstruction is not perfect. The advantage is that the entire collection of 95 measurements has been reduced to just two numbers which convey the information spread over the 95 measurements.

We may summarize the above remarks by saying that the standard deviation is a direct measure of the variation exhibited by the measurements. Earlier we used the *range* as an indication of the spread from the largest to the smallest result. Why abandon such a simple quantity as the range in favor of the standard deviation which required more work to compute?

As we pointed out at the time, to a considerable extent the range also depends on the number of measurements in the collection. Study the following two statements about two operators, each making 50 measurements of the same kind. Operator A, who makes his measurements in sets of ten, has an average range of 18 units, while Operator B, making sets of four measurements, has an average range of 14 units. Which one has the measurements with the wider histogram? Let's consider the problem for a moment. We see that the sets of ten are more likely to include extreme values than the sets of four. Thus if Operator B had selected sets of ten measurements, instead of four, his average range would have been increased by 50 per cent. It would have been 21, not 14, units and he would have the wider histogram. If the range is used — and it sometimes is for quick work — comparisons will be misleading unless the size of the collection is kept constant.

Using the Standard Deviation

The standard deviation does not suffer from the limitation just mentioned for the range, the number of measurements in the collection being automatically allowed for in the formula. Furthermore, the standard deviation uses all the measurements, while the range uses only the two extreme results.

The standard deviation — our estimate of sigma — is a very useful number indeed. First of all, it must never be forgotten that it describes the scatter of *individual* measurements around the average. Suppose our collection consists of 96 measurements. We might divide the 96 values by lot into 24 sets, each with four measurements. Take the average for each set of four measurements. We now have 24 averages. If these averages were given to someone without his being told that they were averages, he might calculate a standard deviation for them.

Can we predict fairly closely the result of this calculation, knowing the value for s that we obtained from the 96 individual readings? The answer is that we can, and by a simple operation. We just divide our estimate, s, for the individual measurements by the square root of four, the number of measurements used in

Table 11. Form for calculating the standard deviation. Data taken from Table 2

mid-value of interval (from Table 2) times 10^4	mid-value of interval minus the average d	number in the interval f	product: difference by number $f \times d$	square of the difference d^2	product: (diff.)^2 by number $f \times d^2$
659.5	−112	1	−112	12544	12544
679.5	− 92	2	−184	8464	16928
699.5	− 72	0	0	5184	0
719.5	− 52	10	−520	2704	27040
739.5	− 32	12	−384	1024	12288
759.5	− 12	18	−216	144	2592
779.5	8	24	192	64	1536
799.5	28	14	392	784	10976
819.5	48	8	384	2304	18432
839.5	68	4	272	4624	18496
859.5	88	2	176	7744	15488
Totals		95	0		136320

$$s = \sqrt{\frac{\Sigma d^2}{n-1}}$$

$$= \sqrt{\frac{136320}{94}}$$

$$= \sqrt{1450} = 38*$$

*Carry square root extraction to two figures.
Note: The sum of the fourth column should be zero. This provides us with a check on the values of d.

each average. Since the square root of four is two, we see that the 24 averages will be spread over about one half the range of values found for the individual measurements. More generally, the average of n measurements will be assigned a standard deviation equal to s divided by the $\sqrt{n}$. Often the standard deviation of an average is given the label *standard error*.

The examples given in the last two chapters showed that no matter how carefully a measurement is repeated, the results obtained in a series of measurements are spread over a range of values. The actual width of the distribution of the measurements varies with different types of measurement, with the care used, and with the quality of the equipment. Intuitively, we conclude that the width of this distribution gives us some clue as to how closely we may have approached the correct value for μ.

Clearly if the measurements stay within a narrow band of values we feel more confidence in our technique of measurement than when the measurements are distributed over a wide range of values. Suppose that a class of seventh grade students made some measurements on the thickness of paper and calculated the standard deviation for their collection of measurements. Also, suppose a senior class in high school made measurements on the thickness of the same paper and calculated their standard deviation. Which class do you think might have the smaller standard deviation? It seems plausible that additional maturity and experience would enable the senior class to make more precise measurements. The histogram for the senior class might be narrower than the one made by the junior class. The magnitude of the standard deviation for each class provides us with mathematical measurement for comparing the two histograms. In fact, if we use the standard deviation, we need not construct histograms to compare the two sets of measurements.

Finding an s for Our Measurements

Now let us undertake to calculate the estimate of the standard deviation of the collection of paper measurements given in Chapter 3. By sorting the actual measurements into eleven classes in Table 3, we have already greatly simplified the number work of finding the average. Sorting into classes makes an even greater saving in arithmetic when calculating s. The arithmetic is shown in Table 11. This short cut gives a numerical result for s that is slightly different from the one obtained by using ungrouped data. The slight difference, however, is of no consequence. In Table 11 the mid-values and the average taken from Table 2 are temporarily multiplied by 10,000. This serves to get rid of the decimals and a lot of zeros right after the decimal point. When we get our answer we simply divide it by 10,000 to put the decimal point back where it belongs. Thus our answer, 38, is really .0038. This temporary multiplying by 10,000 greatly reduces the chance of numerical error.

The arithmetic in Table 11 is a bit time consuming but not difficult. The average, 771.5, is subtracted in turn from the eleven mid-values. The difference is squared and multiplied by f, the number of measurements in the interval. The sum of the squares of these differences is 136320. After substituting in the formula for the standard deviation, this leads to a value for s of 38 or 0.0038 after putting back the decimal point. Remember that s is not σ, the standard deviation, but only the best estimate we can obtain from these data. However, from here to the end of the book we will use the terms s and σ interchangeably.

You are now in a position to test an earlier claim made for the standard deviation — that about two out of three measurements will differ from the average by less than one standard deviation. We found that the 95 measurements of paper thickness had an average value of 0.07715. Now add to and subtract from the average the value 0.0038 which we found for the standard deviation. The result is a lower limit of 0.07335 and an upper limit of 0.08095. These two limits are just one standard deviation away from the average. If you now turn back to Table 2 in Chapter 3, you may count up all the measurements that fall between these two limits. The number of individual values between these limits is 66. Thus 69.5 per cent of the 95 measurements fall between these limits, and this checks in a very satisfactory manner the theoretical percentage of 68.3 (Table 9). Two standard deviations amount to 0.0076. When this quantity is added to and subtracted from the average, we obtain the upper limit, 0.08475 and the lower limit, 0.06955. The per cent of the measurements expected to fall within these limits is 95.4. Therefore we should expect 4.6 per cent of the 95 measurements to be outside these limits. This gives 4.4 measurements. Examination shows that just five measurements in the collection are beyond these 2σ limits.

Although this collection of measurements did not quite fulfill the requirements of complete independence, we find that the standard deviation can be used to predict the number of measurements included in any chosen interval centered on the average.

We take data in the hope that we can get the answer to a scientific question. We want to know whether or not our data provide a satisfactory answer to the scientific question we had in mind. Suppose that someone were doing a project that involved the effect of plant hormones on plant growth. One of the things he might want to know is, how large are the leaves of the treated plants? If it seems worthwhile answering the question, measurements will be made in an attempt to get a satisfactory answer. Of course, one can simply take the average of the measurements and report this. But in general, such an answer is not adequate in scientific investigations.

Student's t

At the very least, it would seem, we should think about what might happen if we repeated the set of measurements. Suppose this were your project and you did repeat the set and got a second average. By this time you are prepared, I trust, to find that there would be some difference between the two averages. What should we do in such a case? You may reply that you would report a grand average based on the two averages. But you should not conceal the fact that there was a difference between the two averages. The simple fact is that if only the first set of measurements had been taken, a repeat of the work will give a different answer. The essential question is "How different?"

It appears that it is not quite enough just to report averages. Something is missing. We would like to make a statement, if possible, that would give some idea of how close our estimate m

has come to the value μ. It would be nice if we could say that our average m does not differ from μ by more than some small amount that we shall call Δ, the Greek letter delta. Now we can not say this and be absolutely sure of being right. We can pick a Δ large enough so that we may have a fairly high degree of confidence that our statement is correct. However, we would like to keep Δ small and still keep our confidence high.

The statement we make has to be based upon the data we have obtained. The indispensable element in making any statement of confidence in our average is s, our estimate of the standard deviation. We have already seen that tables based on the normal law of error make it possible to make statements about the per cent of measurements that fall in an interval centered on the average. Our problem now is somewhat different but closely related. The estimate of σ tells us how much the individual measurements deviate from the average. Our present problem is to make a statement about the average and its closeness to μ. The more measurements in the collection, the better the chance that the average will lie close to μ. There is one way to obtain μ with absolute confidence. That would be to make an infinite number of measurements. Since we will always have some limited number of measurements, the chances that averages of small collections coincide with μ are extremely remote. So let us now study this problem of making a statement about our average that will somehow relate the average to μ.

For a long time many investigators did not attempt to make any statement about the average, particularly if the average was based on very few measurements. The mathematical solution to this problem was first discovered by an Irish chemist who wrote under the pen name of "Student." Student worked for a company that was unwilling to reveal its connection with him lest its competitors discover that Student's work would also be advantageous to them. It now seems extraordinary that the author of this classic paper on measurements was not known for more than twenty years. Eventually it was learned that his real name was William Sealy Gosset (1876–1937).

Suppose we had a collection of n measurements where n may be as small as two. We know how to calculate the average, m, and s, the estimate of σ. We now calculate the quantity, Δ.

$$\Delta = t \; \frac{s}{\sqrt{n}}$$

When Δ is added to and subtracted from the average, this gives an interval which may or may not include within it the unknown μ for which our average, m, is an estimate. We are already familiar with all the symbols in the formula for Δ except for the multiplier t. This number is known as Student's t and is obtained from tables first published by Student.

Suppose we have only four measurements and we desire to have a 50-50 chance that the limits determined by our Δ enclose μ. Student found that in this case the proper value for t is 0.765. If we had eight measurements, the proper value for t would be 0.711. Observe that t is smaller with more measurements as is only reasonable. More measurements give a better average and a better value of s.

Suppose we wish to increase our chance from 50 per cent to a higher probability that our limits include μ. If we want to increase the chance that the interval includes the unknown μ, we must make the interval wider. To raise the probability to 90 per cent (nine chances out of ten), t must be increased to 2.353. Table 12 is a brief table of t. Inspection shows how the value of t depends on a number of measurements and the desired confidence that the interval includes μ.

The first column in Table 12 is headed "Degrees of Freedom" and not "Number of Measurements." Observe that if we have only two measurements each measurement differs from their average by the same amount. If we know one difference, we know that the other difference must be the same.

If there are three measurements, the three differences from their average must sum up to zero if we take the sign into account. Consequently if we are told the values for two of the differences, the third difference is easily found. We can see that the number of independent differences from the average is one less than the number of measurements. The number of independent differences is called *degrees of freedom*.

Table 12. A brief table of *t*.

degrees of freedom	probability			
	.50	.90	.95	.99
1	1.000	6.314	12.706	63.657
2	.816	2.920	4.303	9.925
3	.765	2.353	3.182	5.841
4	.741	2.132	2.776	4.604
5	.727	2.015	2.571	4.032
6	.718	1.943	2.447	3.707
7	.711	1.895	2.365	3.499
15	.691	1.753	2.131	2.947
30	.683	1.697	2.042	2.750
99	.676	1.660	1.984	2.626
∞	674	1.645	1.960	2.576

By calculating Δ we can set limits above and below the average for a set of data with some confidence that μ lies within this interval. The width of the interval enclosed by these limits depends on the value we find for s, on the number of measurements, n, and on the degree of confidence that we wish to have that these limits enclose μ. Naturally every experimenter would like to have these limits as close to m as possible and still have high confidence that μ is included.

The expression

$$\Delta = t \, \frac{s}{\sqrt{n}}$$

shows quantitatively how the limits depend upon the standard deviation and the number of measurements made.

There are two ways by which an experimenter can narrow

these limits. First of all, he may increase the number of measurements. This soon becomes unprofitable because we can see from the formula for Δ that the effect of increasing the number of measurements depends on the square root of the number of measurements. Thus increasing 100 measurements to 200 has a negligible effect. Obviously the most efficient method of narrowing the limits enclosed by plus and minus Δ is to make the numerator, s, smaller. That is the goal of every experiment — and the art of measurement.

With a small set of measurements, say four, we may calculate Δ for the 90 per cent or even the 99 per cent limits. *We have no way of knowing for sure that these limits actually do include μ in any given case.* On the average, i.e., nine times out of ten or 99 times out of 100, such limits will include μ. We will proceed now to try out this technique and see how well it works.

We return to the 95 measurements made on paper thickness. Again we simplify matters by multiplying each value for thickness by 10,000 to get rid of all the decimal points and zeros. Write these 95 values on 95 cards. After shuffling the cards, deal out four of them and consider this a set of four measurements that might have been made. Calculate average m and s for the four measurements. Then calculate for 50 per cent limits

$$\Delta = 0.765 \; \frac{s}{\sqrt{n}} = 0.765 \; \frac{s}{2}$$

Subtract Δ from the set average and add Δ to the set average. We can now say that there is a 50-50 chance that these limits include μ.

What is μ for this investigation? We really don't know, but 95 is a rather large number and, therefore, the average, m, will probably be a close approximation to μ. Thus, it is fairly safe to use m for μ in this experiment. The average is 771 and we may note whether or not the limits we compute do include 771.

Now shuffle the deck again and deal out another four cards and consider these to be another set of measurements. (Be sure to return the first four cards before you shuffle them.) Repeat the computations as before and observe whether or not these new limits for this second set inclose 771. We may continue this process to find out if about half the sets have limits which inclose 771.

You may not want to undertake all this arithmetic, so I have done it for you for 20 sets of four measurements each. In Table 13 I have tabulated for each set its average, its standard deviation, and Δ for 50 per cent limits of confidence, and the limits deter-

Table 13. Calculation of 50 per cent and 90 per cent limits for twenty sets of four measurements drawn by lot from the 95 values in Table 2. All values in Table 2 have been multiplied by ten thousand.

set no.	set ave. m	set S.D. s	$\Delta = 0.765 \frac{s}{\sqrt{4}}$	50% limits	$\Delta = 2.353 \frac{s}{\sqrt{4}}$	90% limits
1	767	7	3	764-770*	8	759-775
2	772	33	13	759-785	39	733-811
3	793	13	5	788-798*	15	778-808*
4	764	14	5	759-769*	16	748-780
5	776	25	10	766-786	29	747-805
6	782	25	10	772-792*	29	753-811
7	780	22	8	772-788*	26	754-806
8	785	26	10	775-795*	31	754-816
9	803	46	18	785-821*	54	749-857
10	774	21	8	766-782	25	749-799
11	784	38	15	769-799	45	739-829
12	753	33	13	740-766*	39	714-792
13	759	20	8	751-767*	24	735-783
14	769	19	7	762-776	22	747-791
15	789	33	13	776-802*	39	750-828
16	794	33	13	781-807*	39	755-833
17	749	57	22	727-771	67	682-816
18	799	23	9	790-808*	27	772-826*
19	774	19	7	767-781	22	752-796
20	762	52	20	742-782	61	701-823

*Values marked with an asterisk did not bracket the grand average of 771.
Note: 0.765 and 2.353 are values of t taken from Table 12.

mined by Δ. An asterisk marks those limits which do not include 771. There are twelve sets with asterisks instead of the expected 10. That is not 50-50, is it? Tossing a coin should give heads or tails on a 50-50 basis. Still we would not be too surprised to get as many as 12 heads (or tails) out of 20 tosses. These 20 sets provide a reasonable verification of our claim.

The last two columns of Table 13 show the 90 per cent limits calculated by using $t = 2.353$. There are only two asterisks in this column marking those sets whose limits did not include the grand average 771. This time we have hit it exactly. Nine times out of ten the limits inclosed the average that was based on a large number of measurements.

We have gone a good way beyond merely reporting an average. We can now attach to an average a pair of limits corresponding to some chosen confidence that the limits will inclose μ. These limits reveal the quality of the measurements and guard us against making undue claims for our average.

Suppose we think of the 95 measurements as constituting one large set out of many large sets of 95 measurements that might have been made. It is easy to set up limits around the average of the 95 measurements. The number of measurements, n, is now 95. Let us select 90 per cent limits. The appropriate value for t may be taken from the next to the last line of Table 12. This is the line for 100 measurements, not 95. But there is very little change in t as n becomes larger. Consequently we may use the value 1.66, to calculate Δ as follows:

$$\Delta = 1.66 \ \frac{38}{\sqrt{95}} = 6.5$$

The value 6.5 is subtracted from and added to 771 to get the limits 764.5 and 777.5. We may convert these back to millimeter units by dividing by 10,000 and obtain the values 0.07645 and 0.07775. The interval between these two limits is a bit more than one thousandth of a millimeter. A statement that the thickness is probably between 0.076 mm. and 0.078 mm. is much more useful than reporting an average of 0.077 mm. without any limits to indicate how uncertain the average may be.

The above computations of Δ were made on the assumption that the 95 measurements were independent. We found previously that some of the girls appeared to have individual biases. A more conservative view of these data would be to consider that we had just 24 measurements, one from each girl. Naturally this would be the average for each girl. You may find it interesting to calculate limits using the 24 averages and taking $n = 24$. The limits will be somewhat wider.

There is still another important way the standard deviation is put to work. Consider two samples of paper that appear to be of much the same thickness. Are they? How could you detect a difference between them and substantiate your claim? Or, how could you say that there is no difference in thickness? Using the technique of measuring a stack of sheets, you could obtain four measurements for each paper. A set of such measurements is shown in Table 14 together with the numerical operations.

Our purpose is to determine whether these data provide substantial evidence of a difference in thickness between the two papers. We must remember that even if we find that the differ-

Table 14. Thickness measurements on two papers; the measurements have been multiplied by 10,000.

	paper A	diff. from ave.	square of diff.	paper B	diff. from ave.	square of diff.
	772	−7	49	765	17	289
	759	−20	400	750	2	4
	795	16	256	724	−24	576
	790	11	121	753	5	25
Total	3116	0	826	2992	0	894
Ave.	779			748		

Average for A minus average for B = 31
Combined sum of squares of the differences = 826 + 894 = 1720

ence between the two papers apparently does not exceed the error of measurement, we still have not proved the papers are equal in thickness. We could make more measurements on each paper. This would give more stable averages and thus might make the evidence for a difference more convincing. On the other hand, there is no point in doing a lot of work if it appears that the difference is too small to be of any practical importance.

Inspection of the two sets of measurements in Table 14 shows that the largest determination of thickness in set B is larger than the smallest in set A, and that the measurements overlap. This suggests that the two papers may, indeed, be of equal thickness. The difference between their averages is 31.

Decisions and Confidence

Our problem now is to determine whether 31 represents a real difference between the averages, or whether it arises simply through the errors in measurement.

If the two samples of paper are of equal thickness, the difference between them would be zero. One way to solve our problem would be to calculate a limit, Δ, with some chosen degree of confidence and see whether zero is included in the range between 31 plus Δ and 31 minus Δ.

The two sets of measurements have been made by the same observer using the same equipment, and therefore should have the same σ. Each set provides an estimate, s, of σ based on four measurements. We will combine the two individual estimates of σ into one estimate in the following manner.

Add together the two sums of squares of the differences to get 1720. Divide this sum by 6. And where does the 6 come from? If we were determining s for each set, our divisor would be 3, $(n - 1)$. Since we are looking for s of the combined measurements we use the sum of the divisors which is 6. So we have $\sqrt{1720 / 6} = 17$. This gives a combined estimate of the standard deviation for this method of measurement. The problem is to find the correct s for the difference between two averages of four measurements. Differences increase the s by the factor $\sqrt{2}$, and averages reduce s by $1/\sqrt{n}$ where n is the number of measurements in each average. The s to use for Δ is

$$s\,\frac{\sqrt{2}}{\sqrt{4}} = 17\,\frac{\sqrt{2}}{\sqrt{4}} = 12$$

We will use this number (already divided by $\sqrt{n}$) to set up limits around the observed difference of 31.

What value of t shall we use? We need values of t for six degrees of freedom, the divisor in our estimate of s. That is, there are three degrees of freedom from each set of four measurements, making a total of six of freedom altogether. So we turn back to Table 12 and select values for t with six degrees of freedom for the 90, 95, and 99 per cent probabilities.

prob.	90%	95%	99%
t	1.943	2.447	3.707
$\Delta = s \times t$	23.3	29.4	44.5
Upper limit	54.3	60.4	75.5
Lower limit	7.7	1.6	−13.5

The limits are obtained by adding $s \times t$ to 31 and subtracting $s \times t$ from 31.

If we pick a probability of 95 per cent we will use a value for t that will, 95 times out of 100, give limits that include the true value of the difference. In this case we use $t = 2.447$ and get the calculated limits around the average of 60.4 and 1.6. We notice that the range 1.6 to 60.4 does not include zero, i.e., no difference between the papers. Thus we can conclude, with 95 per cent confidence, that there is a real difference in thickness between the papers. Strictly speaking, what we have shown is that there is only a small probability (five per cent) that we would obtain a difference as large or larger than 31 if the papers were, in fact, of equal thickness. Consequently we would give up the assumption of equal thickness.

Another investigator might be more cautious about claiming

to have shown a difference between the papers. He elects to work at the more conservative level of 99 per cent probability. In this event his report indicates limits from —13.5 to 75.5. Zero is a possible value between —13.5 and 75.5, so he is unwilling to report that there is a difference at the 99 per cent level of confidence. The choice lies with the investigator, and the importance of the decision greatly influences his choice of a level of confidence. Additional measurements pay off in reducing the standard deviation of the difference between the averages.

There is an extremely important comment to make about this comparison of the thickness of the papers. The *difference* in thickness is not influenced by any bias that may happen to afflict *all* the measurements. Suppose there was a bias $+B$ in the values obtained. This bias will appear in both averages, making each average too large by $+B$. Consequently the difference between the averages is just what would have been found if the measurements had no bias at all. Comparative measurements have this enormous advantage over absolute measurements such as the determination of the gravitation constant, g. Indeed, it is possible to measure and compare the differences between the gravitation constants at two latitudes much more accurately than the constant can be determined at any one latitude.

Measurements and the Work of Scientists

The work of the scientist is tremendously aided by having available very careful measurements on certain standard substances. For example, a chemist who has prepared a new liquid organic compound may be interested in the viscosity of this new substance. Using very simple apparatus, the viscosity can be determined by comparing it with the viscosity of water.

The viscosity of water has been very carefully measured. One way to determine the absolute viscosity is to measure the time required for a given volume of water to flow through a capillary tube of known diameter and length under a given pressure. It is quite an undertaking to establish all these quantities. But if equal volumes of water and the test liquid are compared in identical capillaries under identical conditions of temperature, it is only necessary to measure the time of flow of each liquid and the specific gravity of each liquid. Thus the viscosity of the new compound can be obtained relative to the viscosity of water. When carefully performed, such a comparison is practically free from bias. Of course any bias in the viscosity value assigned to water will be carried over in the value assigned to the new liquid. For this reason the very greatest care is taken in establishing the values of the physical constants for certain reference materials.

What we have been examining is a small part of the theory of measurements. The role of the computations we have just made is to give the investigator an objective basis for making statements about his experimental results.

You may find all this mathematics pretty tiresome and not nearly as much fun as assembling your apparatus and getting it to work properly. And you may ask, "Is it really necessary to go into all these complications?" There are three alternatives, all of which have been widely used in the past and are still used to some extent today. These alternatives are:

1. Arbitrarily make the limits ridiculously wide
2. Guess at the limits
3. Ignore the whole matter of giving your fellow scientists a measure of the quality of your work. That is, just report averages without any limits.

You will agree, I hope, that these alternatives come in a poor second to a piece of work well done and supported by a standard deviation and the narrowest possible limits at the highest possible level of confidence.

6. Instruments

for making

measurements

You may now appreciate why scientists would like to reduce experimental errors in their measurements to the point where they could be ignored. One way to reduce error is to use better instruments. In the paper thickness study we used only a simple metric scale. This may be replaced by a vernier caliper similar to the one shown in Figure 13. The stack of paper can be caught between the jaws of this instrument so there is no need to hold one end of the scale on one edge of the stack. The other advantage — a big one — comes from the auxiliary vernier scale, named after the French mathematician Pierre Vernier (1580–1637).

The auxiliary vernier is a short scale which divides nine divisions on the main scale into ten equal parts. Each whole division of the auxiliary scale equals nine tenths of a main-scale division. The auxiliary scale is used to estimate the tenths of the millimeter. Move the auxiliary-scale zero up to the position on the main scale that is to be estimated. In Figure 14 this position is between 11 and 12 mm. To find the tenths directly, run your eye along the scale until you find a mark on the main scale in line with a mark on the auxiliary scale and read the auxiliary scale. The reading is 11.2 mm. Can you prove that this scheme is sound?

As an example of the usefulness of the vernier caliper, consider the set of measurements made on paper thickness shown in Table 15. The agreement for thickness per sheet is so good in the fourth place that estimates can be given to the fifth decimal place. In this set of measurements the s for the thickness per sheet is 0.00015, one twentieth of the 0.0038 obtained

Table 15. Measurements on paper thickness with a vernier caliper

number of sheets	thickness mm.	thickness per sheet
215	20.0	0.09302
184	17.1	0.09293
146	13.6	0.09315
120	11.2	0.09333
103	9.6	0.09320

Average = 0.09313

by students using a millimeter scale. The vernier caliper makes it much easier to compare the thickness of two papers. Notice, too, how the standard deviation provides a measure of the improvement.

Still further improvement could be obtained with the micrometer, shown in Figure 15. This is a more elaborate vernier and a standard piece of equipment for precision machine work. The shop version reads directly to thousandths of an inch, and tenths of a thousandth may be estimated.

Before we leave the five measurements on paper thickness made with the vernier caliper, let's try another way of looking at them. If a graph is constructed and the actual thickness of each stack is plotted against the number of sheets per stack, the five points should — in theory — lie in a straight line through the origin of the graph. Actually, the points will not lie exactly on a line because of small errors in the measurements. We may ask,

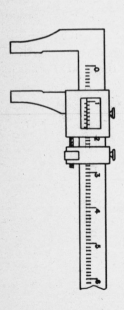

Figures 13 and 14. Objects to be measured with the vernier caliper are held between the jaws. Approximate reading is made on main scale, and readings to nearest tenth are made in auxiliary scale shown in detail at right.

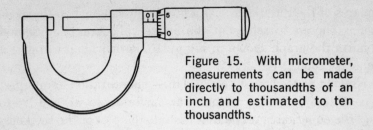

Figure 15. With micrometer, measurements can be made directly to thousandths of an inch and estimated to ten thousandths.

why should we expect them to do so?

The equation for a straight line through the origin is $y = bx$ where b is the *slope* of the line. (Slope is the constant that defines the rise of y values in terms of the increase of x.) The increase in y for one additional sheet on the stack gives the slope. Therefore the slope is an estimate of the thickness of the paper.

The Method of Least Squares

One part of the mathematics of measurement deals with getting the best fit of a line (or curve) to a set of points. What is meant by "best fit"? We can explain fitting a line in terms of the arithmetic average which we accept as the best single number to represent a collection of data. Consider three possible measurements; 13, 8, and 9, represented by the average, 10. The differences between the three measurements and the average are 3, —2, and —1. The sum of the squares of these differences is 14. If any number other than ten is used to get the differences, the sum of their squares becomes larger than 14. The arithmetical average is the number that makes the sum of the squares of the differences a minimum.

Suppose that rather than using the average to represent this set, we use 9, the middle or *median* number. The differences then are 4, —1, and 0. The sum of the squares of these differences is 17. The graph in Figure 16 shows that the sum of the squares of the differences has a minimum value when the average is chosen to represent the entire collection. You should confirm the graph shown in Figure 16. Perhaps you can give a general proof.

Now let us see how we can use this generalization to interpret a collection of measurements. For our example we will try to find the equation of a line that best fits the plot of the thickness of a stack of paper against the number of sheets per stack. Our data will be the vernier readings taken from Table 15. If we let y equal the thickness of the stack in millimeters and x equal the number of sheets per stack, our problem becomes one of considering the various values of b in the equation $y = bx$.

Suppose that we arbitrarily take b as equal to 0.09. Substituting this value in the equation makes it possible to calculate the thickness, y, for any given number of sheets.

x (no. sheets)	calculated y (0.09x)	observed y (thickness in mm.)	difference (obs. y-calc. y)
103	9.27	9.6	+.33
120	10.80	11.2	+.40
146	13.14	13.6	+.46
184	16.56	17.1	+.54
215	19.35	20.0	+.65

All the calculated values for y are below the observed values; without doubt the coefficient b has been given too small a value. The sum of the squares of the differences between observed y and calculated y is 1.1946. By analogy with the arithmetic average, it seems that if we found a value for b that would make the sum of the squares of the differences a minimum, we would have a line that best represents the collection. This procedure is frequently used and it is called the *method of least squares*. We could cut and try various values of b, but this is troublesome. There is a formula that gives the desired value of b directly for lines that pass through the origin.

$$b = \frac{\Sigma\, xy}{\Sigma\, x^2}$$

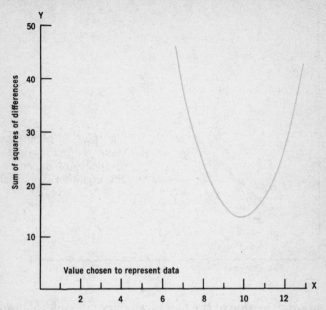

Figure 16. Graph shows Σd^2 when an arbitrary number is subtracted from 13, 9, and 8. Note Σd^2 is a minimum for $x=10$, the average.

Multiply each x by its corresponding observed y and sum up the five products (11764.8). Square each x and sum the five squares (126406). The ratio of these two sums gives b for a line through the origin. The value found for b is 11764.8/126406 = 0.09307. The equation can now be written $y = 0.09307x$. The graph for this equation is shown in Figure 17. The new set of calculated y's and their differences from observed y's show a much improved fit.

x	calculated y (0.09307x)	observed y	difference (obs.-calculated)
103	9.586	9.6	+.014
120	11.168	11.2	+.032
146	13.588	13.6	+.012
184	17.125	17.1	−.025
215	20.010	20.0	−.010

The sum of the squares of these differences has gone down to 0.002089. This sum of squares also leads to an estimate of the

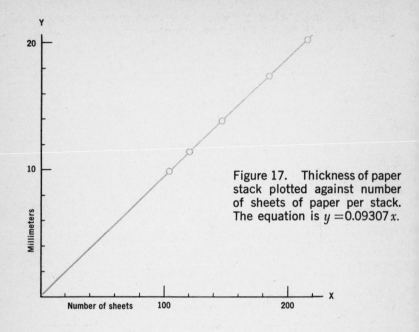

Figure 17. Thickness of paper stack plotted against number of sheets of paper per stack. The equation is $y = 0.09307\,x$.

standard deviation of the measurement of the *width of the stack*. The constant b plays the role of the arithmetic average in our earlier calculations of s. The estimate, s, of the standard deviation is

$$s = \sqrt{\frac{0.002089}{4}} = 0.023 \text{ mm.}$$

Since our measurements were recorded only to the nearest tenth of a millimeter, we probably were rather lucky to get this small s.

If you are sharp eyed you may have noticed that the value 0.09307 for the slope b (the thickness per sheet) is a little less than the average given in Table 15. This is due to the fact that the average in the table gives equal weight to each measurement, while the thicker stacks get more weight in this calculation for the slope.

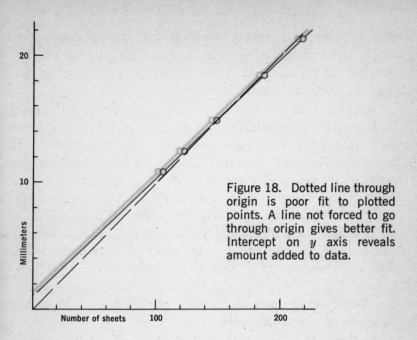

Figure 18. Dotted line through origin is poor fit to plotted points. A line not forced to go through origin gives better fit. Intercept on y axis reveals amount added to data.

(axis labels: Millimeters; Number of sheets; 100; 200; 10; 20)

Which is the better estimate — the average or the slope? The slope is, if we assume that the error of measurement is not influenced by the thickness of the stack. If this assumption is true, the calculations of thickness per sheet should be more reliable for the thicker stacks. The arithmetic average ignores this advantage of the thicker stacks. Usually, however, the discrepancy in the two estimates is unimportant unless the stacks vary greatly in thickness.

As an exercise, try adding one millimeter to all the y values actually observed. That is, imagine that all these measurements are biased by $+1$ mm. The graph and fitted line (the dotted line) for the adjusted data are shown in Figure 18. What is the slope of the line? The line, in an effort to compromise since it *must* go through the origin, runs below the points near the origin and above the points farther out. Even the eye can see that a line not forced to go through the origin would be a better fit to the points. This line has an intercept on the y axis at about one millimeter. An incorrect zero setting for the vernier caliper would make all results high (or low), and thereby introduce a constant error. Perhaps the above discussion will suggest to you one way to detect a constant error or bias.

One major concern in all refined measurements is the calibration of the instrument. Little good would come from carefully reading fractions of a scale division if the instrument itself is in error. The sad news is that errors in the instruments are not revealed by repeated measurements. If a thermometer is in error by half a degree when read at 25° centigrade, this error will not be eliminated by taking the average of many readings all estimated to tenths of a degree. All the readings will have this hidden constant error. There are two ways around this dilemma. One way is to have a thermometer checked by a competent testing laboratory. The laboratory will supply a certificate that gives the corrections to be applied at periodic points along the thermometer scale.

Another way to reveal constant errors is to have one or more similar instruments. One thermometer is used and then replaced by another thermometer. If readings are divided among two or more thermometers, inconsistencies among the thermometers will ultimately be revealed. If two vernier calipers are available, each should be used for half the readings. We may find, just as we did for the magnesium analyses and for the measurements on g discussed in Chapter 4, that there is a difference between the two sets of readings.

Very often this sort of check on instruments can be introduced into experiments without adding appreciably to the labor. If the instruments are in agreement there is, of course, the possibility that both instruments are in error to the same amount and of the same sign. This coincidence is generally regarded as unlikely; so agreement between sets of measurements made by using two or more instruments gives us confidence that constant errors of appreciable magnitude in the instruments are not being overlooked.

One way to learn about instruments at first hand is to make one. We have chosen a simple one, so that you can easily make two or three and compare them. Back in 1887, Captain Andreas Prytz, a Dane, invented a very simple instrument for measuring the area of an irregular plane figure. We all know the formula for the area of a square, rectangle, triangle, and circle. But suppose we need to get the area of a leaf or an irregularly shaped plot of land. The outline could be traced or drawn to scale on graph paper and the number of squares counted.

The measurement of irregular areas is very important in engineering. There are instruments that trace the rise and fall of pressure in an automobile cylinder throughout the piston stroke. The area under the pressure curve must be measured. This information is needed to evaluate the performance of the engine. Civil engineers laying out modern highways use their

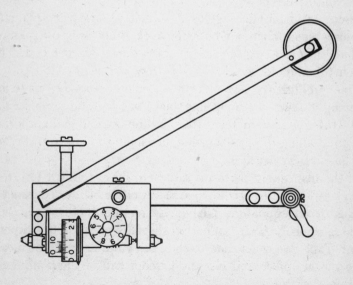

Figure 19.   Irregular areas can be measured with polar planimeter.

transits to determine the profile of a hill. From this profile the engineer will be able to compute the quantity of earth that must be removed. But first, he will need to measure the area enclosed by the profile of the hill. In technical and engineering work irregular areas are measured by a beautiful and expensive instrument called the polar planimeter. See Figure 19.

You will have to make your own planimeter. The hatchet planimeter invented by Captain Prytz can be made very quickly from a piece of coat hanger or other stiff wire. From the lower straight portion of a coat hanger cut a piece about 30 centimeters long. File one end flat and straight across, similar to the base on a right cylinder. File the other end to a tapered but very slightly rounded point so that it will not scratch paper. The blunt end is then hammered out to a hatchet shape. The rough wedge shape should be filed until a sharp arc-shaped edge is formed on the blade as shown in Figure 20.

Now comes the only part requiring care. Bend down about five centimeters of the hatchet end at right angles. Be most particular to have the edge of the hatchet in the same plane as the long piece. Then bend down five centimeters of the pointed end so that it is parallel to the bent-down hatchet end. The hatchet planimeter looks like a very low, very wide letter U.

One preliminary test of this planimeter may be made by drawing it along a straight line. Place hatchet and point on the line. Hold the planimeter vertical by the pointed end and draw the point along the line. The hatchet should follow and stay on the line or very nearly so.

If the instrument performs satisfactorily in the first test, test it on simple figures of known areas. A circle of four-centimeter radius or an equilateral triangle seven centimeters on a side will serve nicely. Graph paper is especially useful in this experiment. Tape two sheets end to end. Draw the figure to be measured on one piece and rest the hatchet end of the planimeter on the other piece.

The measurement is made by picking a point close to the

center of the area to be measured. Draw a straight line from this point to the perimeter. Place the pointed end of the planimeter on the center point of the figure and the hatchet end on the other sheet of graph paper. Generally it is convenient to have the hatchet, the pointed end, and the line to the perimeter all in line and coincident with one of the rulings on the graph paper. Now press the hatchet end down so that it makes a faint but definite mark on the paper. Hold the planimeter upright by the pointed end and move the point along the line from the center out to the perimeter. The hatchet should track along the graph paper ruling. Now trace the outline of the figure and return to the center point. Let the hatchet end follow as it will; do not force it. When the pointed end is back at the center, press down the hatchet to make another indentation in the graph paper.

The area of the figure is computed by multiplying the distance between the two indentations by the distance between the two arms of the planimeter. The hatchet end will have been

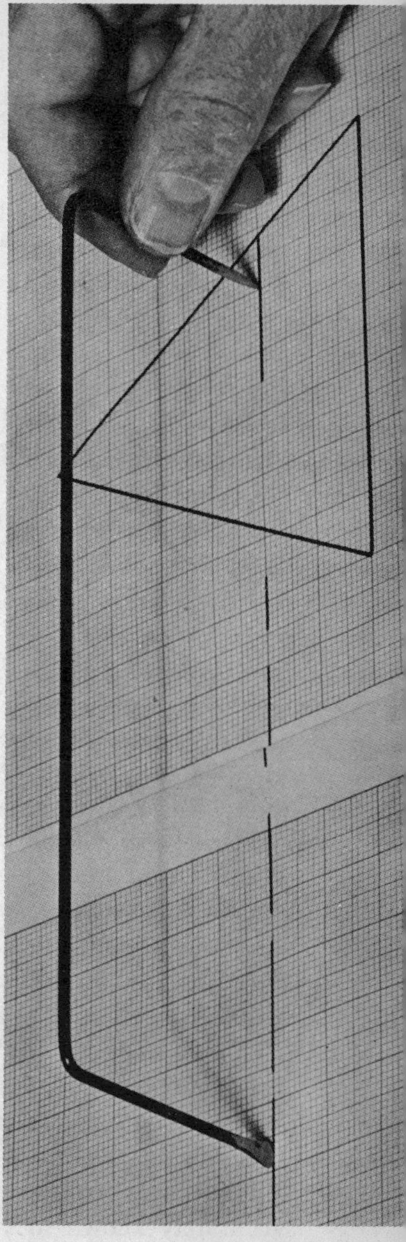

Figure 20. Homemade hatchet planimeter is calibrated by measuring regular figures of known areas.

displaced sideways, i.e., at right angles to its original position. The displacement can be read directly from the graph paper. If the distance between the two arms is also measured by the graph paper, the area can be given in whatever units the graph paper is ruled. The area of a circle or rectangle is also easy to get in square graph paper units.

Repeat the measurement by going around the figure in the opposite direction. You will find the hatchet is displaced in the direction opposite to that in the first trial. The displacements may disagree. Probably this is because hatchet blade and point are not in perfect alignment. Use the average of the two tries as your answer. Repeat such complementary pairs until you have a series of estimates of the area. Be sure to prepare a good form on which to record your data. Determine the error for this method of measuring the area. Refer back to Chapter 5 for the formulas.

One rather satisfying thing about this experiment is that the areas of regular figures drawn on the graph paper are known fairly exactly — much more so than the homemade planimeter measures them. So for all practical purposes, we know the true area. Consequently we can discover if our planimeter has a "constant" error. Does it always miss by about the same amount and of the same sign?

We can in effect "calibrate" this simple instrument. Certainly it would be appropriate to investigate two or more sizes of area and two or more shapes of area. The results of such a calibration study are indispensable in any serious effort to determine the area of some very irregular-shaped area. You can see that mere agreement between repeat tracing on the irregular area does not protect you from some constant error. By testing the instrument and your technique of using it on at least two different, regular figures of known area, you can detect hidden constant errors.

Sometimes it is not easy to have a reference item that is accurately known so you can check yourself. If you make a second planimeter, you will probably find some difference between the sets of measurements made with the two planimeters. The discrepancy between the two sets is a direct warning that there are other errors in addition to those revealed by the scatter of the measurements, all made with the same planimeter. Every investigator who is trying to get very accurate data is confronted with exactly the problems you face in calibrating a homemade planimeter.

We are now at the point where we see that the error in a measurement may be a complex matter. First we studied the variation among repeat measurements. We learned to compute the standard deviation as a measure of the variation among the measurements.

Precision, Accuracy, and Truth

The deviations of the individual measurements from their average determine the *precision* of the measurements. These deviations do not reveal either bias or constant error that may be present in every one of the measurements. Scientists try to arrange their experiments so that the precision, or standard deviation, is the only source of error they need to worry about. This is often achieved by comparing one or more test items with some reference material of known values.

Thus if someone, with much labor, has measured very accurately the thickness of a stack of paper, a sample of this paper may be compared with a paper of unknown thickness. The difference in thickness found between the reference paper and the unknown paper is added to (or substracted from) the value assigned to the reference paper. The only error to consider here is the precision error, since any bias in the measurements does not affect the difference between the two papers.

The scientist who undertakes to establish the correct thickness for the reference paper faces a more difficult problem. Possible biases now become a matter of real concern, and a great deal of effort is required to detect such biases and eliminate them as far as possible. The object is to arrive at an accurate value — one that is close to the true value.

The term *accuracy* involves the error as measured from the *true* value, not as scatter measured around the average of the data. Even if we knew the true value, it is most undesirable to take the differences between the individual measurements and the true value. The differences should always be taken from the average of the data. The t tables only apply when the standard deviation is calculated by using the average. The constant error is revealed by the difference between the average of the measurements and the true value, if one is lucky enough to know the true value. Notice that good precision is required to detect small constant errors.

Table 16. Different values reported for the Astronomical Unit (values 1-12, from *SCIENTIFIC AMERICAN*, April 1961)

number	source of measurement and date	A.U. in millions of miles	experimenter's estimate of spread	
1	· Newcomb, 1895	93.28	93.20	- 93.35
2	Hinks, 1901	92.83	92.79	- 92.87
3	Noteboom, 1921	92.91	92.90	- 92.92
4	Spencer Jones, 1928	92.87	92.82	- 92.91
5	Spencer Jones, 1931	93.00	92.99	- 93.01
6	Witt, 1933	92.91	92.90	- 92.92
7	Adams, 1941	92.84	92.77	- 92.92
8	Brower, 1950	92.977	92.945	- 93.008
9	Rabe, 1950	92.9148	92.9107	- 92.9190
10	Millstone Hill, 1958	92.874	92.873	- 92.875
11	Jodrell Bank, 1959	92.876	92.871	- 92.882
12	S. T. L., 1960	92.9251	92.9166	- 92.9335
13	Jodrell Bank, 1961	92.960	92.958	- 92.962
14	Cal. Tech., 1961	92.956	92.955	- 92.957
15	Soviets, 1961	92.813	92.810	- 92.816

The standard deviation and the constant error should be reported separately. Quite different remedies are required to improve the precision and to reduce constant errors. Both tasks challenge the skill of the experimenter. The experimenter finds it very useful to have some means of demonstrating that improvements in the measurement technique have been achieved. Our discussion is only an introduction to the mathematical treatment of the errors in measurements.

A particularly revealing compilation of measurements made since 1895 of the value of the Astronomical Unit (average distance of the earth from the sun) is shown in Table 16. The table reveals a spread of values reported by the astronomers. This spread refers to the precision of the work and is not a measure of accuracy. The "best" value reported by a later worker is often far outside the limits assigned by an earlier worker.

Make a graph by taking the number of the measurement as x and the reported value as y. The scale on the y-axis should extend from 92.70 to 93.20. You will see that the later values show much better agreement with each other than those in the early part of the century. We have here an impressive demonstration of the increasing refinements of the measurement process. Man is never satisfied. Men will always strive to achieve one more decimal point as they seek to penetrate deeply into the nature of the universe.

7. Experiment with weighing machines

ONE day your author wished to demonstrate the advantage of a carefully planned experiment. There had been some discussion about the accuracy of the weighing machines found in drugstores and other public places. He and three colleagues weighed themselves on four different machines. Each man was weighed on each machine. Each man read his own weight just once. The other three times, his weight was read by his companions. A schedule was set up so that each man's weight was read by all four participants. Each man was weighed on every machine and each man made one reading on each machine. A number of questions could be answered from the 16 weights obtained in this investigation.

First, the weights of the men could be compared. Second, the machines could be compared with one another. Third, the men could be compared with respect to their method of reading a scale. In particular we could discover whether an individual had a tendency to get consistently high or low readings, that is, whether he had a bias.

A schedule was made by dividing a square into 16 smaller squares. Each column of small squares was assigned to one of the men. Each row of squares was assigned to a different machine. The sketch shows the plan at this point.

Man getting weighed

	J.C.	J.Y.	C.D.	M.D.
Machine I				
Machine II				
Machine III				
Machine IV				

Why did we complicate this investigation by having each man read his own weight only once and then have it read the other three times by his companions? This device prevented the second and following readings on a man's weight from being influenced by preceding readings. If a man read his own weight each time, he might have been tempted to make the readings agree a little better with each other than was actually the case. So, as part of the experiment, the readings recorded by any reader were not revealed until the experiment was over. This made sure that we had four readings for any one man's weight by four different readers, none of whom knew what the others had put down.

This foresight, which added nothing to the work, had one other important consequence: if a man had a reading bias and he read his own weight each time, this bias would enter into his own average and into no other average. The difference between the average weights for two men would include their reading biases. An arrangement for each man to read the weights of all four men eliminates this source of error from the comparison of the weights.

The problem was how to enter in the proper box the initials of each individual who was to make the reading. Each man was to read once every man's weight, including his own, and to make one reading on each machine. Clearly each man's initials must appear in all four rows and in all four columns. There are 576 ways in which this can be done. An arrangement of this kind is called a Latin square. The particular Latin square used in this experiment is shown in the next sketch.

Man getting weighed

	J.C.	J.Y.	C.D.	M.D.
Machine I	J.C.	C.D.	M.D.	J.Y.
Machine II	J.Y.	M.D.	C.D.	J.C.
Machine III	C.D.	J.C.	J.Y.	M.D.
Machine IV	M.D.	J.Y.	J.C.	C.D.

The initials in each square designate the man who made the

reading for the particular machine and man associated with that column. We see that J.C. read his own weight on machine I, and read the weight of J.Y. on machine III.

The plan was not yet complete. In what order should the men get on the scales? You may ask, "what difference will that make?" It should not make any difference, is the reply. Whenever possible we try to introduce into our experiments elements that are extremely unlikely to alter the results. This is done as a check. If it turns out that the results are altered by this operation, there may be something wrong with the plan of the experiment, or the data.

How were we to introduce the order of being weighed so as to be perfectly fair to all involved? The obvious way to be fair was to let each man be first once, second once, third once, and last on the scales just once. Furthermore — and this takes a bit of thinking — if a man in his role as reader should read the weight of a first man on the scales, then a second on, a third on, and a last one on, this would further even things out. Therefore, we assigned in each box a number 1, 2, 3, or 4 which told each man listed at the top of each column when it was his turn to get on a particular scale.

The next sketch shows this number entered in the box, and the weight of the man at the top of the column as it was read by the man whose initials are in the box. For example, the numbers tell us that when we came to machine II, C.D. got on the scale first and read his own weight, next J.Y. got on the scale and M.D. read his weight, and so on.

Some of the advantages of this planned experiment appear immediately. The average weight found for each man was a consensus based on all four machines. Since each man was weighed on every machine, the variation among the machines did not enter into the comparison of the weights of the men. That is, if a machine read two pounds too high, the machine introduced this increment for every man. Each man's average would therefore have been increased by half a pound without

	J.C.	J.Y.	C.D.	M.D.	
Machine I	1 J.C. 155.75	3 C.D. 163.25	4 M.D. 153.0	2 J.Y. 169.5	Machine average 160.4
II	4 J.Y. 156.0	2 M.D. 164.25	1 C.D. 153.25	3 J.C. 171.0	161.1
III	2 C.D. 153.0	4 J.C. 161.5	3 J.Y. 151.5	1 M.D. 167.25	158.3
IV	3 M.D. 155.0	1 J.Y. 162.75	2 J.C. 152.25	4 C.D. 169.0	159.8
Man Ave.	154.9	162.9	152.5	169.2	

changing the difference between the average weights for the men. This state of affairs is strictly true only if all the men are approximately the same weight. If one of the men had been replaced by an 80-pound boy, difficulties might have arisen. A machine might have given readings two pounds too high at 160 pounds and also have given readings 1.5 pounds low at 80 pounds. In that case all the men would have gained an extra two pounds and the boy would have lost 1.5 pounds. So the differences between the boy and the men would be altered.

The machines, too, could be fairly compared only in the neighborhood of the average weight of the men. There was a difference of almost three pounds between the averages for machines II and III. The highest weight for each man was given by machine II and the lowest by machine III. Notice that for each machine we form the total weight of all four men. That the men differ in weight does not matter, provided only that they are not too far removed from the over-all average of 160 pounds.

What about the performance of the men as readers of the scales? The scheme was arranged so that each reader took a reading on every machine and for every man. Therefore the same total was put before each reader but in different combinations of machine and man being weighed. The weights read by J.Y. are 169.5, 156.0, 151.5, and 162.75. The total is 639.75 pounds and the average of his four readings is 159.9. Similarly the order on the scales may be examined. The four weights associated with the number 1 include a weighing of each man, a reading by each man, and a weighing from each machine. The weights are 155.75, 153.25, 167.25, and 162.75 and the average is 159.8.

We compute the average for each reader and for the numbers 1, 2, 3, and 4, using the data displayed in the last sketch. All the averages have been assembled together for comparison in Table 17. The men were expected to differ in their weights. The extreme difference between machines was 2.8 pounds. As readers, the men showed excellent agreement; the maximum difference between readers was 0.5 pounds. This difference was just about the same as the 0.4 maximum difference found between the average weight of the men when they were third on and their weight when they were first on the machine. As might be expected, there were small errors in estimating to a quarter of a pound, otherwise the averages for the readers would have agreed exactly.

We cannot explain the differences between machines as arising solely from reading errors. The reading error is clearly quite small by comparison with machine differences. It is safe to conclude that at least some of the machines are slightly in error. The maximum error in a machine, as estimated by the difference between it and the average of all four machines, is 1.5 pounds. This is not enough to worry most users of the machines.

Can we conclude that no machine gives a weight that is off

men		machines		readers		order			
J.C.	154.9	I	160.4	J.C.	160.1	1	159.8	a	160.1
J.Y.	162.9	II	161.1	J.Y.	159.9	2	159.8	b	159.7
C.D.	152.5	III	158.3	C.D.	159.6	3	160.2	c	159.9
M.D.	169.2	IV	159.8	M.D.	159.9	4	159.9	d	159.8
Maximum difference			2.8		0.5		0.4		0.4

by more than 1.5 pounds from the truth? Certainly not. Perhaps the machines were all set by the same mechanic and have a common error. Suppose the machines were purposely all set to read low in order to please those who are concerned about being overweight. And, of course, we have data only in the neighborhood of 160 pounds and no information for much smaller or much larger weights. The easy way to check the machines would be to get hold of some "standard" weights. These are weights that have been checked against the official standards of weight.

The Importance of Experimental Design

In Table 17 there is a fifth column of averages rather mysteriously labeled *a, b, c,* and *d.* Recall that we were able to enter the numerals 1, 2, 3, and 4 in the boxes so that there was a 1 assigned once to each machine, man, and reader. The same holds good for 2, 3, and 4. In a similar way the letters *a, b, c,* and *d* can be put in the boxes so that the letter *a* is assigned once to each machine, man, reader, and numeral.

The averages for the letter *a, b, c,* and *d* ought to agree within the error of reading because no physical action is associated with these letters. The maximum difference for these averages is 0.4 pound. This difference is about the same as the maximum difference found among readers, or for order of getting on the scales. Such a device provides good evidence that the averages for readers agree within the error of reading, and also that the order of getting on the scales did not matter.

The placing of these letters is not shown but is left as an exercise for the reader. Hint: place the a's in the diagonal starting at the top left corner.

When additional symbols are entered in the Latin square, it is then called a Graeco-latin square. Both are widely used in experimental design. A 3 x 3 square permits only four sets of averages; a 5 x 5 square permits six sets of averages. Curiously a 6 x 6 square can be constructed with only *one* set of symbols in the boxes.

Even without formal analysis, the sixteen measurements in Table 17 have revealed a good deal of information. The primary object of the study may be considered to have been a comparison of the men's weights. The same data permitted a check on the weighing machines in regard to possible disagreement among the machines. The data also made it possible to check on possible biases the men might have had as readers. There was no convincing evidence of such biases. The differences among the men as readers were about the same as the small differences associated with the order the men got on the scales, and order should not have had an effect. Most important, the comparison of the men's weights was not impaired by disagreement among the machines. Neither would reading biases have altered the differences found between the weights of the men. Try adding a small bias to any man's readings to verify this statement.

8. Selection of items for measurement

THE problem of getting good measurements and finding ways to describe them concisely has been our chief concern in the preceding chapters. Only indirectly has there been any suggestion that there is sometimes a problem of picking representative items to be subjected to measurement. In Chapter 4, to determine the homogeneity of the magnesium alloy, we tested 50 spots on five bars chosen from 100 bars. Was this a fair sample? We omitted any reference to the question of whether the 923 gasoline pumps could be considered an adequate sample of the total collection of pumps in the country. Taking a large number of items does not guarantee getting a representative selection of the whole supply, or population. Many investigators have found this out the hard way.

Before tackling the general problem of how to get a good selection of items for measurement, let us investigate a special case. Your author went to the bank and got two rolls of newly minted 1961 cents. For the moment, let us accept without argument that these 100 coins give an adequate picture of the coins minted during the work period in which they were made. Undoubtedly the coins accumulated in a big tray and got mixed up in the process. There is a specified weight for cents, together with legal tolerances for minimum and maximum weight of a coin. The nominal weight is 3.110 grams, with a permitted tolerance of 0.130 grams above and below the nominal weight.

The problem was to determine whether or not this sample of 100 pennies fell within the permitted tolerance of 3.110 grams ± 0.130 grams. To do this experiment your author had access to a very fine balance (see Figure 21) that could be read to the *millionth* part of a gram. Such precision was quite unnecessary. The weights were recorded to tenths of a milligram; that is, to the fourth place of decimals. Even that was really unnecessary. Weights to the nearest milligram would have been quite good enough. Why?

The actual weights of the coins vary over a range of about 250 milligrams. Weighing to the nearest milligram would surely be good enough since the weighing error could, at most, extend the range of actual weights by only a milligram or two in 250 milligrams. The weights are correct to four places of decimals. How do we know that? If we reweighed the 100 coins using any other fine balance capable of weighing accurately to six decimal places, we would get exactly the same weights over again out to four decimal places. The possibility of a constant error

Figure 21.   This sensitive balance is capable of accurate
weighings to one millionth of a gram.

common to all weights was eliminated by checking the balance
with a standard weight.

All the above detail is directed to establish that the observed
scatter of the weights is not a result of errors in weighing. The
measurement error is nil for this inquiry. An effort to determine
whether the weights of the coins do or do not conform to specification must depend on examining a number of coins. For this

purpose we need a balance good enough so that a coin will not be called outside the tolerance limit because our weighing introduces additional scatter into the results. There is no measurement error in the weights as recorded. The variation among the weights found is a property of the coins, and in no wise reflects measurement error.

class interval grams	number of coins in		total for 100 coins
	1st 50	2nd 50	
2.9800 - 2.9999	0	1	1
3.0000 - 3.0199	1	3	4
3.0200 - 3.0399	3	1	4
3.0400 - 3.0599	2	2	4
3.0600 - 3.0799	5	2	7
3.0800 - 3.0999	9	8	17
3.1000 - 3.1199	13	11	24
3.1200 - 3.1399	8	9	17
3.1400 - 3.1599	6	7	13
3.1600 - 3.1799	3	3	6
3.1800 - 3.1999	0	2	2
3.2000 - 3.2199	0	1	1
Total	50	50	100

Table 18. Weights of new U.S. cents weighed on an accurate balance

The individual weights are given in class intervals of 20 milligrams for each roll of coins. These are displayed in Table 18. A glance at the totals in the last column suggests that the actual weights of the coins are distributed among the class intervals in very much the same manner as are the measurement errors on one object such as the paper thickness measurements.

When the coins themselves were arranged in columns corresponding to the class intervals, they formed a histogram that is indistinguishable from those exhibited in Chapter 2. We have here, not 100 crude measurements on one object, but one very careful measurement on each of 100 objects. Nevertheless the 100 results are distributed in the same form as the normal law of error.

The same calculations that were made on repeated measurements on one object are appropriate for this collection of single measurements on each of 100 objects. A standard deviation may be calculated and the same probability statements made that were explained in Chapter 5. So there is no difficulty in arriving at a concise description of this collection of weighings. The

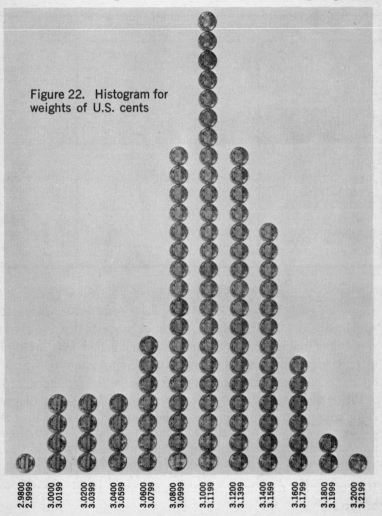

Figure 22. Histogram for weights of U.S. cents

2.9800
2.9999

3.0000
3.0199

3.0200
3.0399

3.0400
3.0599

3.0600
3.0799

3.0800
3.0999

3.1000
3.1199

3.1200
3.1399

3.1400
3.1599

3.1600
3.1799

3.1800
3.1999

3.2000
3.2199

reader may calculate the average and standard deviation. Compare the average with the nominal weight, 3.110.

Some may wonder what would be the result if these coins were weighed on a rather crude balance. If the balance is such that repeated weighing of the same object gave a standard deviation of more than 20 milligrams, the spread of the distribution would be substantially increased. If s_c is the standard deviation of the coins, and s_b the standard deviation for weighings made with the crude balance, then the standard deviation of the observed weights for coins weighed with the crude balance is easily calculated. It is $\sqrt{s^2{}_c + s^2{}_b}$. If duplicate weighings are made on each of, say, 50 coins, it is an easy matter to solve for both s_b and s_c. We can calculate what the standard deviation for coins would be if they were weighed on an errorless balance.

We delayed answering the question as to whether these 100 coins are completely representative of the total population of newly minted cents. The chances are very good, of course, that the coins are representative of some rather short interval in the total annual production of cents. If we desire to report on the cents minted in any year, we should plan to put aside a few coins from each day's work. These coins should be thoroughly mixed and a sufficient number drawn from the collection. The coins are enough alike so that there is virtually no danger of bias or intentional selection, either for those going into the collection or for those drawn for weighing.

Coins are easy to set aside and mix thoroughly. However, there are many cases in which it is physically impossible to do this. For example, bales of raw wool weigh several hundred pounds, and there may be hundreds in a shipment. Custom duties are assessed on the basis of the per cent of clean wool in the shipment. Since the bales vary considerably, it becomes necessary to sample the shipment by taking two or more cores from each of a number of bales. Economy of effort requires that no more bales be sampled than is necessary to obtain a satisfactory estimate of the per cent of clean wool in the entire shipment. We have already learned one way to sample fairly. The bales may be marked 1 to n. Then n cards are numbered 1 to n. If k bales must be sampled, k cards are drawn from the carefully shuffled pack.

Random Digits

There is a more convenient way to attain a random selection without using a pack of numbered cards. You can use random number tables, just page after page filled with the digits 0 to 9 put down in random order. Do not think this random order is easy to get. Oddly enough, very special efforts have to be made to avoid excessive inequalities in the frequencies of the digits and to avoid recurring patterns. A very small selection of 1600 random digits is given in Table 19.

A table of random numbers can be used in various ways. Suppose we wish to draw 12 bales from a shipment of 87 bales. The bales are first numbered 1 to 87. Now go down a column of paired random numbers. Start at the top left of Table 19 and write down pairs of digits. Omit pairs 88, 89, . . . , 90, and 00, because there are no bales corresponding to these numbers. We find the pairs 44, 84, 82, 50, 83, 40, 33, (50), 55, 59, 48, 66, and 68. The second 50 is omitted because that bale is already in the sample. These are the bales to be sampled. This method avoids any deliberate attempt to put good, or poor, bales in the sample. One may start at any part of the table so there is no possibility of anyone's influencing the selection. In fact every effort must be made to start in different places.

If there were 302 bales, we would take the digits in triplets. Can you see the triplet, 441, in the two numbers 44 and 17? Any number from 001 up to and including 302 gets in the sample. There will be many numbers greater than 302. If a number is greater than 302 and less than 605, subtract 302 from it to get your number. If the number is 605 up to 901, subtract 604 to get your number. Ignore numbers above 907. In this way

Table 19. Random Numbers

```
44 17 16 58 09    79 83 86 19 62    06 76 50 03 10    55 23 64 05 05
84 16 07 44 99    83 11 46 32 24    20 14 85 88 45    10 93 72 88 71
82 97 77 77 81    07 45 32 14 08    32 98 94 07 72    93 85 79 10 75
50 92 26 11 97    00 56 76 31 38    80 22 02 53 53    86 60 42 04 53
83 39 50 08 30    42 34 07 96 88    54 42 06 87 98    35 85 29 48 39

40 33 20 38 26    13 89 51 03 74    17 76 37 13 04    07 74 21 19 30
96 83 50 87 75    97 12 25 93 47    70 33 24 03 54    97 77 46 44 80
88 42 95 45 72    16 64 36 16 00    04 43 18 66 79    94 77 24 21 90
33 27 14 34 09    45 59 34 68 49    12 72 07 34 45    99 27 72 95 14
50 27 89 87 19    20 15 37 00 49    52 85 66 60 44    38 68 88 11 80

55 74 30 77 40    44 22 78 84 26    04 33 46 09 52    68 07 97 06 57
59 29 97 68 60    71 91 38 67 54    13 58 18 24 76    15 54 55 95 52
48 55 90 65 72    96 57 69 36 10    96 46 92 42 45    97 60 49 04 91
66 37 32 20 30    77 84 57 03 29    10 45 65 04 26    11 04 96 67 24
68 49 69 10 82    53 75 91 93 30    34 25 20 57 27    40 48 73 51 92

83 62 64 11 12    67 19 00 71 74    60 47 21 29 68    02 02 37 03 31
06 09 19 74 66    02 94 37 34 02    76 70 90 30 86    38 45 94 30 38
33 32 51 26 38    79 78 45 04 91    16 92 53 56 16    02 75 50 95 98
42 38 97 01 50    87 75 66 81 41    40 01 74 91 62    48 51 84 08 32
96 44 33 49 13    34 86 82 53 91    00 52 43 48 85    27 55 26 89 62

64 05 71 95 86    11 05 65 09 68    76 83 20 37 90    57 16 00 11 66
75 73 88 05 90    52 27 41 14 86    22 98 12 22 08    07 52 74 95 80
33 96 02 75 19    07 60 62 93 55    59 33 82 43 90    49 37 38 44 59
97 51 40 14 02    04 02 33 31 08    39 54 16 49 36    47 95 93 13 30
15 06 15 93 20    01 90 10 75 06    40 78 78 89 62    02 67 74 17 33

22 35 85 15 13    92 03 51 59 77    59 56 78 06 83    52 91 05 70 74
09 98 42 99 64    61 71 62 99 15    06 51 29 16 93    58 05 77 09 51
54 87 66 47 54    73 32 08 11 12    44 95 92 63 16    29 56 24 29 48
58 37 78 80 70    42 10 50 67 42    32 17 55 85 74    94 44 67 16 94
87 59 36 22 41    26 78 63 06 55    13 08 27 01 50    15 29 39 39 43

71 41 61 50 72    12 41 94 96 26    44 95 27 36 99    02 96 74 30 83
23 52 23 33 12    96 93 02 18 39    07 02 18 36 07    25 99 32 70 23
31 04 49 69 96    10 47 48 45 88    13 41 43 89 20    97 17 14 49 17
31 99 73 68 68    35 81 33 03 76    24 30 12 48 60    18 99 10 72 34
94 58 28 41 36    45 37 59 03 09    90 35 57 29 12    82 62 54 65 60

98 80 33 00 91    09 77 93 19 82    74 94 80 04 04    45 07 31 66 49
73 81 53 94 79    33 62 46 86 28    08 31 54 46 31    53 94 13 38 47
73 82 97 22 21    05 03 27 24 83    72 89 44 05 60    35 80 39 94 88
22 95 75 42 49    39 32 82 22 49    02 48 07 70 37    16 04 61 67 87
39 00 03 06 90    55 85 78 38 36    94 37 30 69 32    90 89 00 76 33
```

every number from 1 to 302 gets an equal chance to be drawn. Earlier in the book you were told to use cards. I must now confess that, instead of cards, I used a table of random numbers because it was less work. Besides, cards stick together.

Most of you are aware that the census taker puts a lot of questions to some people, while others are asked for only a little information. Modern measurement theory shows that properly drawn small samples are quite satisfactory measures of the whole population. Often these small samples are even better than a complete count because they can be made by a few well-trained census takers who will make fewer mistakes. In all such samples, the use of random selection is absolutely indispensable to avoid various sources of bias in the results. This subject of sampling is so vast that many large volumes have been written on it in recent years.

9. Measurement

of thread strength

Tₕᵢₛ chapter describes a method of measuring the strength of sewing thread. The results will not add to our store of factual knowledge, but we will become further acquainted with the problem of making a measurement. We are going to start pretty nearly from scratch and make an apparatus out of easily obtained parts. I am sure you will discover just how exasperating a piece of apparatus can be. This elementary problem drives home the fact that making measurements is not child's play.

We are going to load a piece of thread until it breaks. On way to do this would be to tie one end of the thread to a suppor and attach a light container to the other end. Pour in sand ver slowly and stop as soon as the thread breaks. Now weigh th container and contents and you have your answer. You can d the experiment this way if you like. Of course you will repea the experiment many times in order to reveal the scatter of th results. What do you think causes the scatter? Is it the difficult in making the measurement, or do different specimens actuall have different strengths?

Making a Testing Apparatus

The method we are going to use is more complicated. Th reason for making it more complicated is that the apparatu involves features found in actual test apparatus. The scheme i one that multiplies the weight applied to the thread by mean of a lever. This permits testing relatively strong material wit moderate loads. Our test device will double the load applied The illustration (Figure 23) shows a stick, 52 cm. long with 13 cm. crossarm mounted on a wooden base. The pieces ar joined by small right-angle brackets with the upright brace with a guy wire. A hook is screwed in the underside of th crossarm about nine cm. out from the upright. The specimen t be tested hangs from this hook. A small eyelet is screwed i near the base of the upright. Finally ten small brads are place close together in a row along the edge of the baseboard and on more brad about 20 cm. away from the middle brad in th group of ten.

Our lever is made from a short piece of coat hanger wire A piece 16 cm. long will do. File a shallow notch half a cm. from each end. Take care to keep the notches in line. Now turn the wire over and file another shallow notch halfway between the end notches. This notch should be 180° from the other two

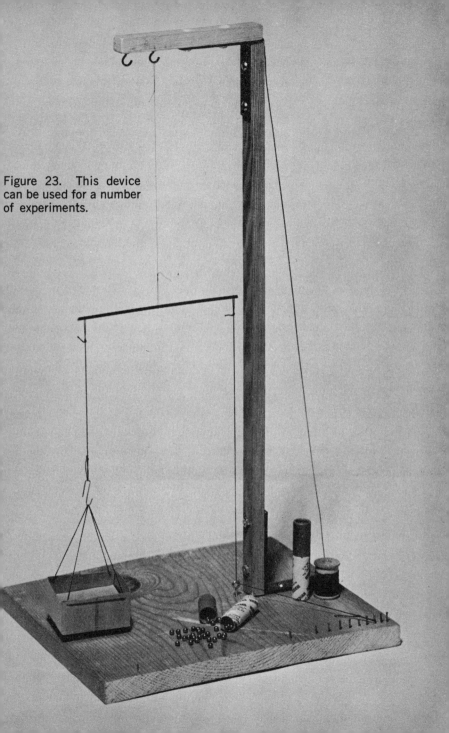

Figure 23. This device can be used for a number of experiments.

A small cardboard box 5 x 7 cm. x 3 cm. deep can be put together with Scotch tape. This allows layers of six nickels to be placed in the box. Eight layers for a total of 48 nickels were used in the experiment. Strong linen thread loops are attached to the box. On top of the nickels rests a false cardboard bottom. BB shot are then added one by one until the load breaks the specimen.

We are now ready to assemble the apparatus. All thread, other than the test specimen, is strong linen thread. The test specimen, 30 cm. in length, is taken from a spool of size A mercerized cotton thread. A loop is tied in one end of the test thread and the loop placed over the middle brad in the row of brads. The other end is doubled around the lone brad in the corner of the board. Hold the thread together, slip the thread off the brad, and tie the loop in that end of the specimen. This standardizes the length of the test specimen at about 20 cm.

Take a piece of linen thread and tie loops in each end so that the length between the tips of the loops is about 40 cm. A similar piece about 14 cm. between loop tips will also be needed.

Hang the test specimen from the hook. Slide the notched wire lever through the free loop and settle the loop in the center notch. In the notch in the end of the wire near the upright, hang a loop of the long linen thread. Lead the free end through the eyelet and place the loop over one of the brads so that the lever is a little high on the free end. When the load is applied to the other end, the test specimen stretches and the free end of the lever will get lower. In the notch on the free end of the lever, hang a loop of the short linen thread. In the other end attach a paper clip with an end bent out to make a hook. Hang the box of nickels on this hook.

We are now ready to test the apparatus to see if it works properly. Add BB shot one by one until the specimen breaks. It should break between the loops. A break at a loop may result if the notch has a rough edge. The box should clear the base by about one cm. Too big a drop causes the BB's to jump out and perhaps be lost. Place a spool against the side of the box to prevent rotation.

BB Units

The end of the lever near the upright may be regarded as a fulcrum. The load is applied twice as far from the fulcrum as the point of support by the specimen. Hence the load on the specimen is doubled. The load includes the box and box contents and hook. The specimen supports the wire, but the weight of the wire is not doubled. Why? After you have adjusted the apparatus and acquired some skill in using it, you can proceed to measure a number of specimens.

Nickels are used as the "standard" of weight because the nominal weight of a nickel is five grams. The lever, supported by a linen thread in the center, was used as a balance to find the weight of the box, of the lever, and the conversion factor for nickels to BB's.

Several nickels were attached to a thread by cellulose tape and hung on one end of the lever. The box and accessories (but without nickels) were hung on the other end. BB's were added until box and contents balanced the nickels. Some more nickels were added and the number of BB's were again increased to get balance. From these data it was easy to get the weight of the box in BB's (20) and the conversion factor of 14 BB's to one nickel.

How could we get the weight of the lever? Easy. We cut another piece of wire the same length, notched it, and weighed it against box and BB's. The weight was found to be 18 BB's. All measurements were reported in BB's.

We have rather casually taken for granted that we managed to get the middle notch exactly midway between the two end notches. If we did not, then the factor two will not give the load on the specimen. We will more than double the load if the shorter end is near the upright, and do less than double the load if the longer end is near the upright. The two ends of the lever were marked I and II so that a record could be made of whichever end was near the upright. A series of tests could be run with the lever in one position and a second series with the lever reversed. Unequal division of the lever would ultimately be revealed by the averages for the two series.

In thinking about this problem a bit, it appears that if two trials are run — reversing the lever on the second trial — the effect of unequal division virtually cancels out when the average of the two trials is taken. The trials were run in pairs and the results for seven pairs of tests are given in Table 20. Examination of the 14 results shows considerable variation among the specimens. The seven results with end I near the support may be compared with the seven made with end II near the support. The technique described in Chapter 5 is appropriate.

The average breaking strength of the thread in grams is obtained by dividing the number of BB's by 14 to get them converted to nickels. Then multiply by five to get grams. Why bother with BB's at all? Why not just use nickels? A nickel is a pretty big weight to drop in the box, and the shock effect would break the thread prematurely. The BB's are also a convenient way of estimating fractions of a nickel.

If the lever had been divided into quite unequal parts we would expect one member of a pair to give consistently higher results than the other member. End I is the higher four times, and II has the higher result three times. The high variability of the thread obscures the slight inequality of division. If the result with end I near the support is divided by the total for the pair, we get the seven ratios: .492, .543, .529, .542, .473, .499, and .557. The average of these is .5193. A t test could be run to see if the limits for this average include the value 0.5000. If so, the evidence would be insufficient to show an unequal division of the lever. You should also satisfy yourself that the way to calculate the position of the middle notch is to take the result with end I near the support and divide by the total for the pair.

Table 20. Measurement of thread strength. All weights in BB's
Weight: Lever = 18; Box = 20; 48 nickels = 672

expt. no.	end at support	BB's added	box plus nickels	load total	twice load	lever wt.	total BB's	ave. BB's
1	I	150	692	842	1684	18	1702	
	II	177	692	869	1738	18	1756	1729
2	I	257	692	949	1898	18	1916	
	II	105	692	797	1594	18	1612	1764
3	I	247	692	939	1878	18	1896	
	II	144	692	836	1672	18	1690	1793
4	I	222	692	914	1828	18	1846	
	II	80	692	772	1544	18	1562	1704
5	I	104	692	796	1592	18	1610	
	II	196	692	888	1776	18	1794	1702
6	I	183	692	875	1750	18	1768	
	II	185	692	877	1754	18	1772	1770
7	I	299	692	991	1982	18	2000	
	II	93	692	785	1570	18	1588	1794

This will give the distance from end I to the middle notch as a fraction of the distance between the two end notches.

Aside from the fun of assembling the apparatus, we have seen another way in which a "constant" error can enter into measurements. If the middle point is not exactly halfway between the end notches, a bias is introduced. We have seen how a proper program of work (reversing the lever) puts us in a position where we can correct for the bias easily and automatically. The data also make it possible to "test" the lever for bias at the same time you are collecting data to determine thread strength. We have also obtained some idea of how the thread strength varies.

The experiment could be easily extended in scope. We used a standard length for the test specimen. Suppose we used a specimen twice or half as long, do you think this would influence the results? Think carefully. You could run the tests by permitting the box to rotate and compare with tests in which the box is not allowed to rotate. You could try the effect on the strength of boiling the thread, or exposing it to direct sunlight for two weeks. You could compare different brands of thread or different colors.

The apparatus and the technique of using it are there to serve whatever line of inquiry interests you. An interesting line of research soon diverts attention from the measurement problem itself. Nevertheless the problem of measurement is still there. Although it hardly needs to be said, men have found that good research depends on good measurement.

I wish that I could assure you that the vast scope and variety of measurement problems have been revealed in this book. Instead, I must warn you that there is much more to the subject. Not all measurements are best described by the normal law of error. There are many special situations and distributions that apply to counts of radioactive particles, to seed germination, and to opinion polls. Other distributions are used for measurements on the fatigue failure of metal parts, and still others apply in the study of flood heights and in reliability studies.

There are, however, elements common to all these distributions. Consider the notion of unexplained variation. Apparently there is something subtle in the notion of random numbers and in using random procedures of selection. Because these concepts of variability and randomness are common to all measurement problems, they have been the major objects of our attention.

A large part of our attention has been given to how measurements are obtained. Yet, the science of measurement, like a coin, has two sides to it. One side shows the ingenuity and skill of experimenters in devising better methods of measurement for their individual researches. The other side of the coin deals with the properties common to all measurements. While this side of the coin has dominated our discussions, the two sides are inseparable.

As scientists explore the unknown regions beyond the present frontiers of science, they encounter new problems and new kinds of measurements. Often these data pose entirely new problems to the measurement specialist. Working together, scientist and measurement specialist push back the frontiers of knowledge; still the frontiers grow longer. One thing appears certain — there is no limit to the length of the frontiers of knowledge.

Table of Squares

	0	1	2	3	4	5	6	7	8	9
0	0	1	4	9	16	25	36	49	64	81
1	100	121	144	169	196	225	256	289	324	361
2	400	441	484	529	576	625	676	729	784	841
3	900	961	1024	1089	1156	1225	1296	1369	1444	1521
4	1600	1681	1764	1849	1936	2025	2116	2209	2304	2401
5	2500	2601	2704	2809	2916	3025	3136	3249	3364	3481
6	3600	3721	3844	3969	4096	4225	4356	4489	4624	4761
7	4900	5041	5184	5329	5476	5625	5776	5929	6084	6241
8	6400	6561	6724	6889	7056	7225	7396	7569	7744	7921
9	8100	8281	8464	8649	8836	9025	9216	9409	9604	9801
10	10000	10201	10404	10609	10816	11025	11236	11449	11664	11881
11	12100	12321	12544	12769	12996	13225	13456	13689	13924	14161
12	14400	14641	14884	15129	15376	15625	15876	16129	16384	16641
13	16900	17161	17424	17689	17956	18225	18496	18769	19044	19321
14	19600	19881	20164	20449	20736	21025	21316	21609	21904	22201
15	22500	22801	23104	23409	23716	24025	24336	24649	24964	25281
16	25600	25921	26244	26569	26896	27225	27556	27889	28224	28561
17	28900	29241	29584	29929	30276	30625	30976	31329	31684	32041
18	32400	32761	33124	33489	33856	34225	34596	34969	35344	35721
19	36100	36481	36864	37249	37636	38025	38416	38809	39204	39601
20	40000	40401	40804	41209	41616	42025	42436	42849	43264	43681
21	44100	44521	44944	45369	45796	46225	46656	47089	47524	47961
22	48400	48841	49284	49729	50176	50625	51076	51529	51984	52441
23	52900	53361	53824	54289	54756	55225	55696	56169	56644	57121
24	57600	58081	58564	59049	59536	60025	60516	61009	61504	62001
25	62500	63001	63504	64009	64516	65025	65536	66049	66564	67081
26	67600	68121	68644	69169	69696	70225	70756	71289	71824	72361
27	72900	73441	73984	74529	75076	75625	76176	76729	77284	77841
28	78400	78961	79524	80089	80656	81225	81796	82369	82944	83521
29	84100	84681	85264	85849	86436	87025	87616	88209	88804	89401
30	90000	90601	91204	91809	92416	93025	93636	94249	94864	95481
31	96100	96721	97344	97969	98596	99225	99856			

SELECTED READINGS

A Million Random Digits, by the Rand Corporation. Glencoe, Ill., The Free Press, 1955. Dip into it at random.

Design of Experiments, by R. A. Fisher. New York, Hafner Publishing Company, 1949. Read the first 100 pages.

Facts from Figures, by M. J. Moroney. New York, Penguin Books, Inc., 1958. Easy to read and inexpensive.

Introduction to Statistical Analysis, by W. J. Dixon and F. J. Massey, Jr. New York, McGraw-Hill Book Company, Inc., 1957. A widely used college textbook.

Mathematical Recreations and Essays, by W. W. Rouse Ball and H. S. M. Coxeter. New York, The MacMillan Company, 1947. Read about Euler's officer's problem.

Probability and Statistics, by F. Mosteller, R. E. K. Rourke, and G. B. Thomas, Jr. Reading, Mass., Addison-Wesley Publishing Company, Inc., 1961. Textbook for Continental Classroom on Television.

Statistical Methods for Chemists, by W. J. Youden. New York, John Wiley and Sons, Inc., 1955. A small book, not limited to chemists.

Statistics, by L. H. C. Tippett. New York, The Home University Library of Modern Knowledge. Oxford University Press, 1944. Very elementary.

Statistics: A New Approach, by W. Allen Wallis and H. V. Roberts. Glencoe, Ill., The Free Press, 1956. A lively introduction by professionals for amateurs.

"Student's" Collected Papers, edited by E. S. Pearson and John Wishart. London, The Biometrika Office, University College, 1947. Wise essays on making measurements.

Technological Applications of Statistics, by L. H. C. Tippett. New York, John Wiley and Sons, Inc., 1950. Chapter 8 gives an account of the density measurements on nitrogen that led to the discovery of the rare gases.

GLOSSARY

Accuracy   Refers to the discrepancy between the true value and the result obtained by measurement.

Average   Refers to arithmetic average, m, or mean. If n measurements have been made, the average is obtained by dividing the sum of all n measurements by n.

Average deviation   If n measurements have the average m, the sum of the deviations (ignoring signs) divided by n gives the average deviation.

Bias   Refers to a more or less persistent tendency for the measurements, as a group, to be too large or too small.

Deviation   The difference between a measurement and some value, such as the average, calculated from the data.

Class interval   An arbitrarily selected interval into which measurements are grouped on the basis of their magnitude.

Error   In the study of measurements "error" does not mean "mistake," but is a technical term denoting deviations from the average or some other computed quantity. Such deviations are considered to be *random errors*. Bias involves the notion of a *constant error*.

Estimate   A numerical value calculated from data. The average is an estimate of the quantity under measurement. Other parameters such as the standard deviation, σ, are often estimated from the data.

Graduation mark   The marks that define the scale intervals on a measuring instrument are known as graduation marks.

Histogram   A graphical representation of a collection of measurements. Equal intervals are marked off on the x axis. A rectangle is erected on each interval, making the heights of the rectangles proportional to the number of measurements in each interval.

Least squares   A mathematical procedure for estimating a parameter from a collection of data by making the sum of the squares of the deviations a minimum.

124

Normal law of error   A mathematical equation that in many cases describes the scatter of a collection of measurements around the average for the collection.

Parameter   A parameter is the property or quantity that the measurements are expected to evaluate. The word parameter is used for the correct value of the property.

Precision   Refers to the agreement among repeated measurements of the same quantity.

Population   Refers to a group of items belonging to a well-defined class from which items are taken for measurement.

Random   A random procedure for selecting items from a population gives every member of the population equal opportunity to be chosen.

Range   The difference between the largest and smallest values in a collection of measurements.

Standard deviation   Estimated from n measurements by calculating

$$s = \sqrt{\frac{\Sigma(\text{dev})^2}{n-1}}$$

where $\Sigma(\text{dev})^2$ means the sum of the squared deviations from the average.

Standard error   Sometimes used for the standard deviation of an average. It is equal to the standard deviation divided by the square root of the number of measurements used to get the average.

Symbols

Σ = Capital sigma, summation sign

μ = population average, true value

m = arithmetic average of the measurements, an estimate of μ

σ = population standard deviation

s = estimate of σ computed from data

e = base of natural logarithms, a mathematical constant

π = ratio of circumference to diameter of a circle, a mathematical constant

b = conventional symbol for slope of a straight line

t = Student's t, a multiplying factor for s used to obtain probability limits about the average of a collection of data.

Tolerance   An agreed-upon permissible departure from specification.

Unit   Every measurement is expressed as a multiple or fraction of some appropriate, well-defined unit quantity such as centimeter, volt, etc.

Vernier   A mechanical aid for estimating fractions of a scale interval.

INDEX

W. J. YOUDEN

Among the small number of recognized experts in the field of statistical design of experiments, W. J. Youden's name is likely to come up whenever the subject is mentioned. A native of Australia, Dr. Youden came to the United States at a very early age, did his graduate work in chemical engineering at the University of Rochester, and holds a Doctorate in Chemistry from Columbia University. For almost a quarter of a century he worked with the Boyce-Thompson Institute for Plant Research, where his first interest in statistical design of experiments began. An operations analyst in bombing accuracy for the Air Force overseas in World War II, he is now statistical consultant with the National Bureau of Standards, where his major interest is the design and interpretation of experiments. Dr. Youden has over 100 publications, many in the fields of applied statistics, is author of *Statistical Methods for Chemists*, and is in constant demand as a speaker, teacher, columnist, and consultant on statistical problems.